U.S. Marine Corps Combat Water Survival

Plus

Marine Scouting and Patrolling

Department of Defense

Pentagon
Publishing

MCRP 3-02C

Marine Combat Water Survival

U.S. Marine Corps

PCN 144 000069 00

To Our Readers

Changes: Readers of this publication are encouraged to submit suggestions and changes that will improve it. Recommendations may be sent directly to Commanding General, Marine Corps Combat Development Command, Doctrine Division (C 42), 3300 Russell Road, Suite 318A, Quantico, VA 22134-5021 or by fax to 703-784-2917 (DSN 278-2917) or by E-mail to **morgann@mccdc.usmc.mil**. Recommendations should include the following information:

- Location of change
  - Publication number and title
  - Current page number
  - Paragraph number (if applicable)
  - Line number
  - Figure or table number (if applicable)
- Nature of change
  - Add, delete
  - Proposed new text, preferably double-spaced and typewritten
- Justification and/or source of change

Additional copies: A printed copy of this publication may be obtained from Marine Corps Logistics Base, Albany, GA 31704-5001, by following the instructions in MCBul 5600, *Marine Corps Doctrinal Publications Status.* An electronic copy may be obtained from the Doctrine Division, MCCDC, world wide web home page which is found at the following universal reference locator: **http://www.doctrine.usmc.mil**.

Unless otherwise stated, whenever the masculine gender is used, both men and women are included.

DEPARTMENT OF THE NAVY
Headquarters United States Marine Corps
Washington, D.C. 20380-1775

6 January 2003

FOREWORD

Marine Corps Reference Publication (MCRP) 3-02C, *Marine Combat Water Survival*, provides Marine Corps combat water survival techniques, procedures, and training standards. This publication also teaches Marines to cross water obstacles and perform water rescues correctly and safely.

This publication is the foundation for teaching Marines correct water survival techniques and procedures that are used throughout the Marine combat water survival program (MCWSP). Once an individual or a unit has completed the MCWSP, this publication can be used as a refresher course before water operations.

MCRP 3-02C supersedes Fleet Marine Force Reference Publication (FMFRP) 0-13, *Marine Combat Water Survival*, dated 16 September 1991.

Reviewed and approved this date.

BY DIRECTION OF THE COMMANDANT OF THE MARINE CORPS

EDWARD HANLON, JR.
Lieutenant General, U.S. Marine Corps
Commanding General
Marine Corps Combat Development Command

Publication Control Number: 144 000069 00

MCRP 3-02C Marine Combat Water Survival

Table of Contents

Chapter 1. Survival at Sea

Abandoning Ship 1-2
Jettisoning Equipment 1-2
Abandoning Ship Technique 1-3
Modified Abandoning Ship Technique 1-6
Surface Burning Oil Swim 1-9
Surviving With a Pack 1-11
Preparing Equipment 1-12
Tying Waterproof/Plastic Bags 1-13
Packing the Pack 1-13
Swimming With the Pack 1-13
Staying Afloat With a Life Preserver 1-15
Inherently Buoyant Life Preservers 1-15
Inflatable Life Preservers 1-15
Staying Afloat Without a Life Preserver 1-18
Floating With an Inflated Blouse 1-18
Floating With Inflated Trousers 1-20
Sling Method 1-20
Splash Method 1-26
Blow Method 1-32
Avoiding Heat Loss in Cold Water 1-36
Individual Protection From the Cold 1-37
Group Protection From the Cold 1-38
Drownproofing Methods 1-39
Crawl Stroke 1-40
T-Method 1-44
The Sweep 1-46

Breast Stroke........1-48
Side Stroke........1-52
Elementary Backstroke........1-57

Chapter 2. Water Rescues

Reaching Rescue Techniques........2-2
Reach........2-2
Reach From a Deck........2-3
Arm Extension........2-3
Leg Extension........2-4
Wading Assist........2-4
Throw........2-5
Lifesaving Approaches........2-7
Front Surface Approach........2-7
Rear Approach........2-7
Approach Strokes........2-8
Crawl Stroke Approach Stroke........2-8
Breast Stroke Approach Stroke........2-11
Level Offs........2-13
Front Surface Approach........2-13
Single Armpit Level Off........2-14
Double Armpit Level Off........2-16
Rescue Techniques........2-18
Wrist Tow........2-18
Single Armpit Tow........2-19
Double Armpit Tow........2-21
Collar Tow........2-22
Cross-Chest Carry........2-22
Tired Swimmer's Assist........2-24
Defense Against a Drowning Victim........2-25
Block........2-25
Wrist-Grip Escape/Wrist-Grip Escape Alternative.......2-26

Front Head-Hold Escape 2-27
Rear Head-Hold Escape 2-29
Administering First Aid/Rescue Breathing 2-31

Chapter 3. Treatment of Casualties and Avoidance of Dangerous Marine Life

Drowning 3-1
Symptoms 3-1
Treatment 3-2
Hypothermia 3-2
Symptoms 3-4
Survival Time 3-4
Treatment 3-4
Heat-Related Injuries 3-6
Burns 3-7
Common Medical Problems Associated with Sea Survival 3-8
Seasickness 3-8
Saltwater Sores 3-8
Blindness/Headache 3-9
Constipation 3-9
Sunburn and Dehydration 3-9
Dangerous Marine Life 3-10
Sharks and Barracudas 3-11
Sea Snakes 3-12
Poisonous Fish 3-12
Turtles and Moray Eels 3-12
Corals 3-12
Jellyfish, Portuguese Man-of-War, Anemones, and Others 3-12
Spiny Fish, Urchins, Stingrays, and Cone Shells 3-13

Chapter 4. Negotiating Water Obstacles

Tides....4-1
Surf....4-2
Plunging Waves....4-3
Spilling Waves....4-4
Surging Waves....4-4
Currents....4-4
Offshore Currents....4-4
Rip Currents....4-5
Littoral Currents....4-5
Back Bays....4-6
Rivers and Canals....4-6

Chapter 5. Fording Waterways

Selection of a Ford Site....5-1
Determine the Slope....5-2
Clinometer....5-2
Map....5-3
Line of Sight and Pace....5-3
Determine the Current Speed....5-4
Measure River Width....5-5
Calculate Downstream Drift....5-6
The Buddy System....5-7
Water Crossings....5-8
Care of Weapons....5-8
ISOMAT Raft....5-9
Poncho Raft....5-10
Construction of a Pack Raft....5-12
Single-Rope Bridge....5-13
High and Dry Crossings....5-17

Swift Current Crossings 5-18
Slow Current Crossings 5-19
Removal 5-19

Appendix. Knot Tying

Square Knot Appendix-1
Bowline Appendix-2
Hitches Appendix-2
Half Hitch Appendix-2
Two Half Hitches Appendix-3
Round Turn and Two Half Hitches Appendix-3
Butterfly Knot Appendix-4

Throughout history, water has posed special challenges to Marines and Sailors during times of both peace and war. Therefore, the inherent dangers associated with waterborne operations demand that Marines and Sailors receive proper water survival training. Combat units that are confident in their ability to work in and around water can use the water to their advantage in combat, and history is filled with examples of how the proper preparation or training for survival in water reduced or averted disaster.

USS *Indianapolis*
On Sunday, 29 July 29 1945, the heavy cruiser USS *Indianapolis* (the *Indy*) was en route to the Philippine Sea. Shortly before midnight (about 39 hours out of port), the *Indy*, running blacked out and unescorted, was rocked by two explosions on her starboard side. With communications smashed, the ship could not signal its distress and sank within 15 minutes.

Three life rafts and a floater net supported a few survivors, but the rest of the survivors drifted about, held up by rubber life belts or Mae Wests. About 60 seamen died the first night. Survivors assumed the ship would be reported overdue in Leyte, and that they would be rescued within 2 days. By Monday evening, however, panic began to set in as some life jackets lost their buoyancy from the long immersion. Some men even fought over life jackets, which resulted in at least 25 deaths. No one dared sleep for fear of losing his jacket.

Throughout the next several days, in-transit aircraft flew nearby without spotting the desperate seamen. As best they could, the men kept together, some tying long ropes to each other, floating like corks on a net.

Not until late Thursday morning, 3 1/2 days after the ship sank, were the men discovered. Luckily, a plane on a routine flight over the area spotted the survivors. When ships picked them up that night, the survivors learned they had never been reported overdue. Every one of the 1,196-man crew was a casualty; 880 crewmembers were listed as dead or missing. Although many lives were lost, the innovative and expedient use of flotation devices and float techniques employed by the survivors helped save hundreds of lives.

USS *America*

On Thanksgiving eve, 23 November 1995, the USS *America* made its way through the Arabian Sea. Twenty-year-old, Marine Lance Corporal (LCpl) Zachary Mayo was unable to sleep and, wanting some fresh air, made his way onto an open-air platform near the aircraft hangar bay, which was three levels below the sleeping quarters. While he was on the platform, the ship veered suddenly, throwing LCpl Mayo through the platform's protective bars and into the sea, 30 feet below.

Frantic, LCpl Mayo called out in vain to the watchmen on the flight deck, which was 64 feet above him. It soon became clear to him that the USS *America* would keep its course into the Gulf of Oman until his absence was discovered at morning muster. The LCpl took a moment to consider his situation. Since land was at least 100 miles away, swimming was suicide; he would have to stay afloat until a search party found him.

Using the techniques he had learned during combat water survival training, LCpl Mayo made a flotation device out of his coveralls and tried to relax. Meanwhile, business continued as usual aboard the USS *America.* Since LCpl Mayo was on special assignment with the hazardous materials division, his absence went unnoticed until a petty officer asked several Sailors if they had seen their shipmate recently. By the time a roll call had been completed, LCpl Mayo had been adrift at sea for over 24 hours. Although three, fixed-wing Viking aircraft were deployed to search for LCpl Mayo, most people aboard ship feared the worst.

After 34 hours at sea, LCpl Mayo was discovered by fishermen on a Pakistani fishing boat. LCpl Mayo's survival is a testament not only to his incredible physical courage, but also to the soundness of the lifesaving training and techniques he received during combat water survival training.

Chapter 1
Survival at Sea

As a Marine, you face a variety of potential water emergencies whenever you cross expanses of water: ships, watercraft, and amphibious assault vehicles (AAVs) can sink; aircraft can crash into the sea; or you can accidentally fall into the water. However, there are some basic precautionary measures you can take to protect your safety and reduce your chances of becoming a water casualty. Determine the following information as soon as you board any type of vessel. Your knowing the following information may save your life or the life of your fellow shipmates.

- How many life preservers and lifeboats/rafts are on board?
- Where are the life preservers and lifeboats/rafts located?
- What type of unit survival equipment is on board?
- Are individual survival kits issued to each person on board?
- How much food, water, and medicine do the survival kits contain? When was the last time the contents were inspected for proper quantities and shelf life expiration?
- Is there sufficient survival equipment available for the number of personnel?
- How many other personnel are there on board, and where are they located?
- What are the egress procedures for the ship, boat, watercraft, AAV, or aircraft?

Abandoning Ship

When you embark on a Navy ship, you will receive abandoning ship instructions from Navy personnel. If given the order to abandon ship, report to your designated assembly area and put on a life preserver. DO NOT inflate the life preserver until you are clear of the ship. Torn life preservers will not inflate and inflated life preservers can block you, and those behind you, from exiting the ship. A flotation device that has been inflated may also burst if you jump from a significant height. See pages 1-15 through 1-35 for staying afloat with and without a life preserver.

DO NOT remove your clothing, boots, or shoes before abandoning ship. Your trousers and blouse may be the only flotation devices available if your life preserver is faulty or becomes damaged, and your clothes can provide some insulation from the cold water. However, remove your soft cover and place it in a cargo pocket for later use. The soft cover is both lightweight and a good protection against sunburn caused by the sun's rays reflecting off the water.

Jettisoning Equipment

Equipment should be kept properly packed and waterproofed in case you have to abandon ship. If entering the water from a height greater than 30 feet, wearing your equipment (e.g., pack, helmet, gas mask) could cause injury. Upon impact with the water, the helmet will "cup" air inside of it. The chin strap may also create a "hanging effect" as you submerge from the force of the fall. Therefore, you should remove your helmet and gas mask before abandoning ship.

If you are unable to maintain buoyancy due to the amount of equipment secured to your pack and body, then jettisoning some of your equipment may become necessary. Equipment that you should always retain includes canteens of freshwater, first aid kit, soft cover, and survival kit. The survival kit should include first aid items, water purification tablets or drops, fire starting equipment, signalling items (e.g., flashlight, strobe light, chemlights), food procurement items, and shelter items. Other items in a survival kit include sunburn lotion and lip balm, knife, goggles/sunglasses, plastic bag, matches and lighter, mirror. See Marine Corps Reference Publication (MCRP) 3-02F, *Survival,* for a detailed list of survival items and applications.

Abandoning Ship Technique

When abandoning ship, safety considerations must be observed. Use the following technique when abandoning ship without your combat gear:

Place your hands on their opposite shoulders, forming a crisscross pattern.

Step to the edge of the ship's deck and check the water below for debris or survivors. If the water is clear, look straight ahead and prepare to jump. If the water is not clear, move to another location.

> NOTE: DO NOT hold your nose as you abandon ship. If you do hold your nose, the force of impact into the water could jar your arm and hand and cause you to break your nose.

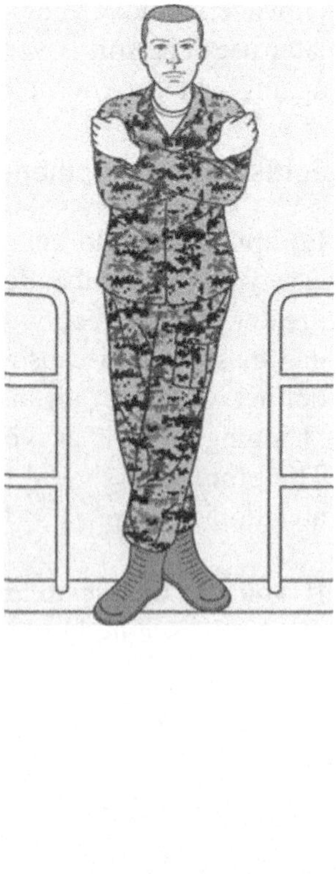

Step off the side of the ship with a smooth, 30-inch stride. DO NOT DIVE OFF THE SHIP. DO NOT LOOK DOWN AT THE WATER. LOOK STRAIGHT AHEAD. Looking down at the water can render you unconscious or cause injuries upon impact.

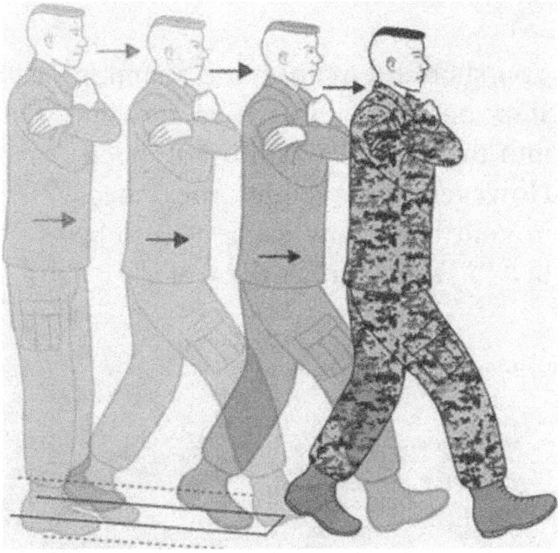

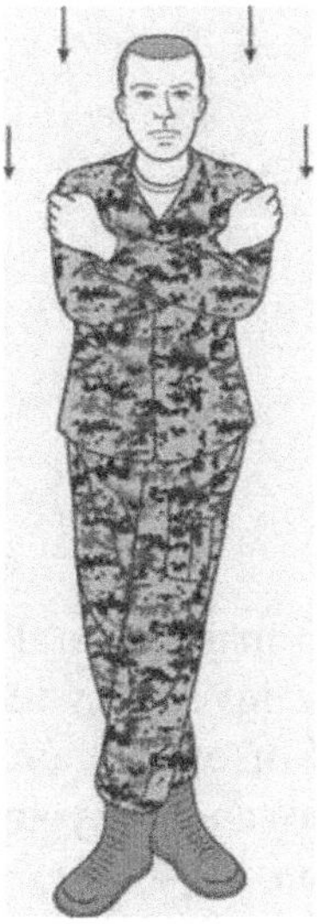

Bring your trailing leg forward during the fall. Cross your trailing leg behind your leading leg.

Keep your head parallel to the water's surface until hitting the water.

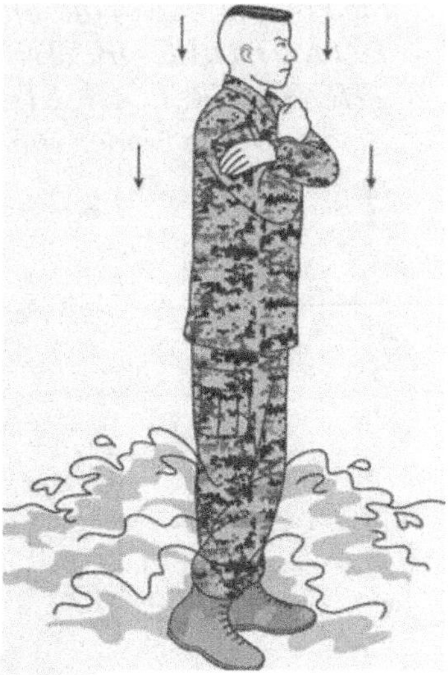

You should remain in the abandon ship position until your descent into the water has almost stopped. However, the weight imbalances in your body may cause you to be in a "J" shape under the water

.

Once your downward motion has ceased, your feet may be parallel with the ocean bottom or you may be nearly inverted with your feet over your head. To counteract potential disorientation, you should pause briefly and allow the natural buoyancy of your torso to bring your body to a nearly upright position.

Floating debris can cause hazards. Therefore, you should swim upward, extending one arm (hand is shaped as a fist) upward to feel for obstructions. If you encounter debris, try to push it away or surface in a different location.

Swim away from the ship. DO NOT LOOK BACK AT THE SHIP. Looking back could slow your movement away from the area. Remember, your objective is to leave the area as quickly as possible because—

- Equipment and debris may be falling from or spilling out of the ship.
- Additional casualties can occur if individuals abandoning the ship fall on top of swimmers already in the water.
- Swimmers close to the sinking ship may get pulled underneath the water by the suctioning effect of the ship as it goes under.

Modified Abandoning Ship Technique

When abandoning ship while wearing full combat gear (weapon, helmet, and a properly waterproofed pack), safety considerations must be observed. The modified abandoning ship technique is used while wearing full combat gear and exiting from a height less than 30 feet.

Warning

When exiting from a height greater than 30 feet, remove your helmet and pack. Fasten the helmet to the pack or place it inside the pack before jettisoning. If jettisoning gear from the ship or aircraft, check the water below for survivors before throwing the gear forward of the intended jump area. Once in the water, retrieve your gear and swim out of the area.

Place your weapon over one shoulder, muzzle down, with the weapon parallel to your side.

Place your arm and hand along the weapon and hold it to your side.

Take your free hand and place it on top of your helmet to prevent neck/spinal injury from the force of the water pulling upward on the helmet as your body enters the water.

Note: An alternate method is to place your weapon over one shoulder, muzzle down, with the weapon parallel to your side. Reach across your body and grasp the sling of your weapon and hold it to your body. Take your free hand and place it on top of your helmet.

Step to the edge of the platform and check the water below for debris and survivors. If the water is clear, look straight ahead and prepare to jump. If the water is not clear, move to another location.

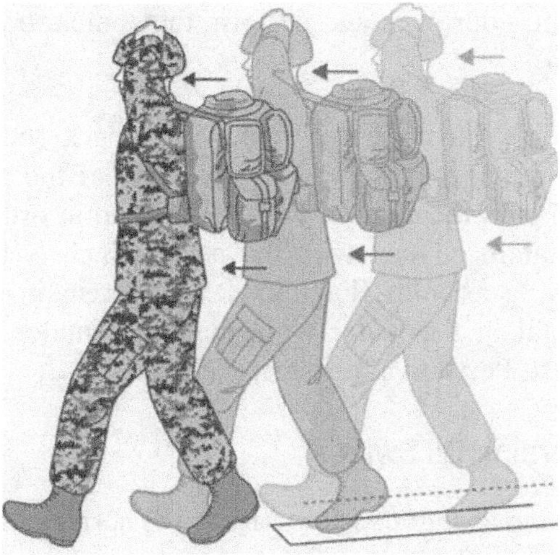

Step off the side of the platform with a smooth, 30-inch stride. DO NOT DIVE OFF THE SHIP, DO NOT LOOK DOWN AT THE WATER, LOOK STRAIGHT AHEAD. Looking down at the water can render you unconscious or cause injury upon impact.

Bring your trailing leg forward during the fall. Cross your trailing leg behind your leading leg.

Keep your head parallel to the water's surface until hitting the water.

You should remain in the modified abandoning ship position until your descent into the water has almost stopped. The buoyancy of a properly waterproofed pack will immediately pull you to the surface. Once you break the water's surface, unsling your weapon and loop the sling over your head. You may have to seesaw the sling so as to ensure the sling passes between your helmet and pack. Once over your head, loop the sling around your neck with the weapon aligned in the center of your body, muzzle down. Lean back on your pack and perform the combat travel stroke to exit the area.

> *NOTE:* You may also remove the pack (unsnap the quick-release strap on one side of the pack, and pass the strap to your other hand in order to maintain contact with the pack, which is functioning as your flotation device). Remain horizontal in the water with your pack under your chest. Perform a breast stroke kick.

Surface Burning Oil Swim

After you have abandoned ship, rise to the surface using the techniques shown on the page 1-6. However, you must remember that fuel from sinking ships or downed aircraft will float on the surface of the water. Therefore, you must move clear of the floating fuel by swimming away from the ship or aircraft as soon as possible. Either swim upwind (into the wind) of the ship/aircraft or swim against the current. Either method allows you to move away from the fuel and the wind/current will push the fuel past you. To properly execute a surface burning oil swim—

Extend your arms overhead as far as possible.

Wave your arms back and forth vigorously to splash a hole while moving upward.

Splash as long as possible to push burning fuel away from the surfacing area.

Use your arms and hands to sweep away fuel and debris.

Kick your legs in a constant breast stroke kick.

Extend your arms (palms outward) forward on the surface, arms shoulder-width apart.

Pull your hands in and back toward the chest.

Stop your hands in front of your face and rotate them so that your palms face forward (roughly halfway out of the water).

Sweep your arms forward to a full extension at the shoulder width. This splashes debris, oil, or burning liquids aside. To reduce the chance of fatigue, use two short splashes to the front to extend the path.

Repeat the preceding step as necessary while swimming clear of the area.

Surviving With a Pack

If packed properly, your pack will float, and it is your key piece of equipment for staying afloat and overcoming water obstacles. If the pack's contents are properly waterproofed, it can support you (with a combat load) in the water. Buoyed up by a waterproofed pack, you eventually emerge from the water with all your equipment (e.g., boots, helmet, flak jacket, weapon, survival items).

Your pack floats based on a scientific principle known as Archimedes' principle. This principle states that an object submerged in a liquid is buoyed up by a force equal to the weight of the liquid displaced (pushed aside) by the object. If the weight of the displaced liquid is greater than the weight of the object, the object floats. If the weight of the displaced liquid is less than the weight of the object, the object sinks. For example, a machine gun sinks in the water, but it still weighs less in the water than it does on land. Even though a machine gun sinks, it is still buoyed up by a force equal to the weight of the water it displaces. For this reason, you should not try to hold yourself or your equipment any higher out of the water than they would naturally float; doing so wastes both energy and body heat.

Preparing Equipment

Before packing your pack, you must prepare your gear/equipment. Tape or pad all sharp edges and equipment corners. Ideally, your gear/equipment is placed in plastic bags and the plastic bag is then placed inside the standard-issued, rubberized, waterproof bag. Waterproof bags are not completely water tight; henceforth, the added protection of first wrapping the gear/equipment in a plastic bag, then placing the plastic bag inside the waterproof bag. Large plastic bags (e.g., trash bags) work well for bulky equipment (e.g., sleeping bags, field jackets, shelter halves, gas masks). Small plastic bags work best for small items (e.g., shaving gear).

> *NOTE:* If a gas mask must be carried outside the pack, cover it with a waterproof bag.

You will need the following items, which are available through the supply system, to prepare your equipment for packing:

- Waterproof bags:
  Pistol bag, plastic size #1 (8" x 18").
  Rifle bag, plastic size #2 (10" x 56").
  Machine gun bag, plastic size #3 (15" x 56").
  Multipurpose bag, cover, plastic size #4 (20" x 84").

- Panama work vest (life preserver).
- Small plastic bag, self sealing (6" x 6" and 12" x 12").
- Riggers tape, olive drab (2 1/2").
- Willie peter bag, waterproof clothing.

Tying Waterproof/Plastic Bags

Try to remove the excess air from the waterproof/plastic bag before securing the bag's opening. If filled with air, the bag can burst if pressed from the outside. Perform the following steps to tie a waterproof/plastic bag:

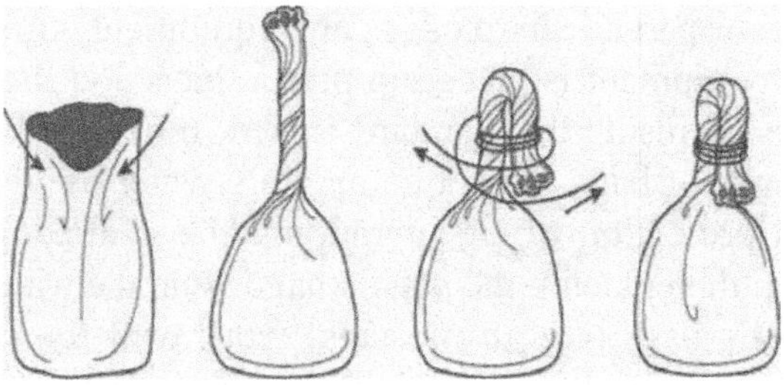

Packing the Pack

Place filled and tied bags inside the pack. Carefully handle sharp items (e.g., tent pegs, poles) to prevent puncturing the bag. Place items in the pack in order of expected use. Close the pack and its compartments. Attach sleeping mats or bags as high as possible on the outside of the pack.

Swimming With the Pack

A useful technique for propelling yourself forward while wearing a pack is the combat travel stroke. This technique allows you to float nearly horizontal and to propel yourself forward with bicycle-style kicks and breast stroke-style arm sweeps. To execute the combat travel stroke—

Body position
Keep the upper part of your body prone to the water, let your legs dangle horizontally, and keep your face up.

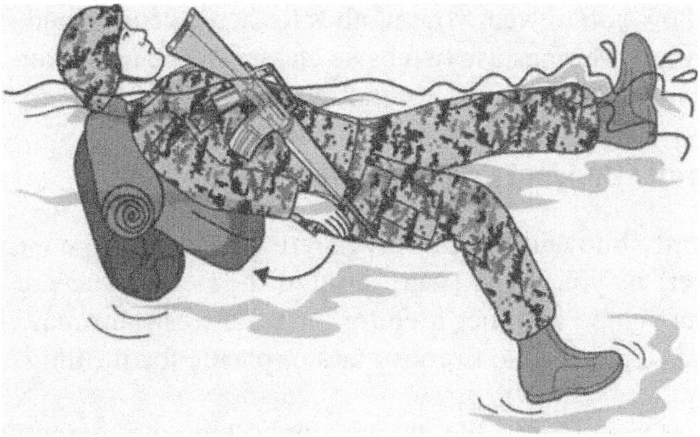

Arm action
Extend your hands out in front of your waist. Sweep your arms slightly downward and back 90 degrees to propel your body through the water. Move your hands back to the front of your waist. Repeat.

Leg action
Continuously move your legs in a bicycle-like movement, bringing your knees up high and step out.

Breathing
Keep your face out of the water during the stroke, and breathe freely.

Staying Afloat With a Life Preserver

The best form of flotation is to find any kind of floating object that will keep you and your equipment out of the water or minimize your exposure to the water. Life preservers are the best method, as they allow you to wear your clothes for heat retention and sunburn prevention. Marines use two basic classes of life preservers: inherently buoyant life preservers and inflatable life preservers.

Inherently Buoyant Life Preservers

Inherently buoyant life preservers are either vest-type (worn like a jacket) or yoke-type (worn around the neck). The preserver's outer envelope is either a cotton or water resistant material that encloses a removable fibrous glass or plastic foam filling.

In the Marine Corps, the most common type of inherently buoyant life preserver is the vest-type with collar, known as the kapock preserver. The kapock consists of collar straps, upper front chest straps, leg straps, and waist drawstrings that secure the preserver to you. The leg straps, which are fitted on both sides of the life preserver, ensure that the preserver remains around your chest while you are in the water. A chest strap is attached to the life preserver to facilitate lifting you out of the water. The strap can also be attached to other survivors or to lifeboats to reduce the fatigue that results from holding onto a floating/secured object by hand.

Inflatable Life Preservers

Marine Corps aircraft and AAVs have inflatable life preservers on board. The inflatable life preserver issued to Marines is known as the LPP (life preserver personal). The LPP is capable of both oral inflation and CO_2 cartridge inflation. The LPP consists of buoyancy chambers, CO_2 inflator, and an oral inflation tube. The buoyancy chamber is deck gray in color and is made from a neoprene-coated nylon fabric.

Inflatable life preservers must be stored in a cool, dry place. Heat, moisture, and light cause deterioration of the life preserver material. Do not stow CO_2 cylinders near steam lines or radiators. Heat can increase the pressure inside the cylinders causing them to explode. Avoid sharp edges in stowage. Sharp edges increase wear and tear on the life preservers and may also puncture inflatable buoyancy chambers.

> **CAUTION:** Do not inflate the life preserver until you are clear of the aircraft, ship, or vehicle. Torn life preservers will not inflate and inflated life preservers can block you, and those behind you, from exiting the aircraft, ship, or vehicle.

To don and adjust the LPP—

Remove the LPP from its storage container.

Fasten the belt fasteners in front with the pouch in the rear.

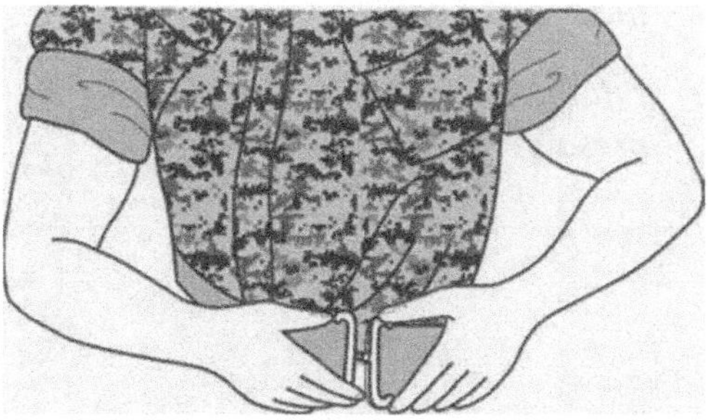

Adjust the belt to fit; secure any excess belt by mating the hook and pile tape.

Rotate the pouch to the front and re-adjust the belt if necessary.

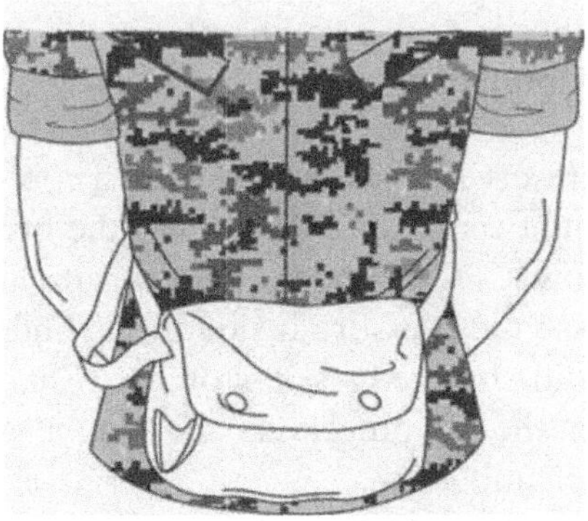

Open the snap fasteners on the pouch and unfold the life preserver.

Place the deflated life preserver over your head.

Place the storage container into the pouch after donning the life preserver.

Lift the lower end of the life preserver out of the pouch.

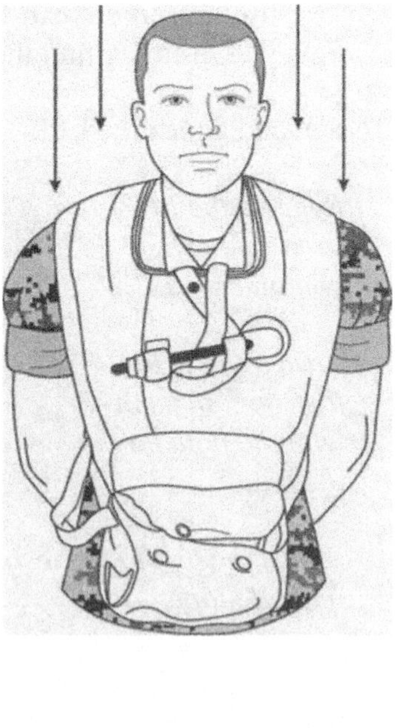

Inflate the life preserver by pulling on the lanyard attached to the CO_2 inflation valve or by blowing on the end of the oral inflation valve.

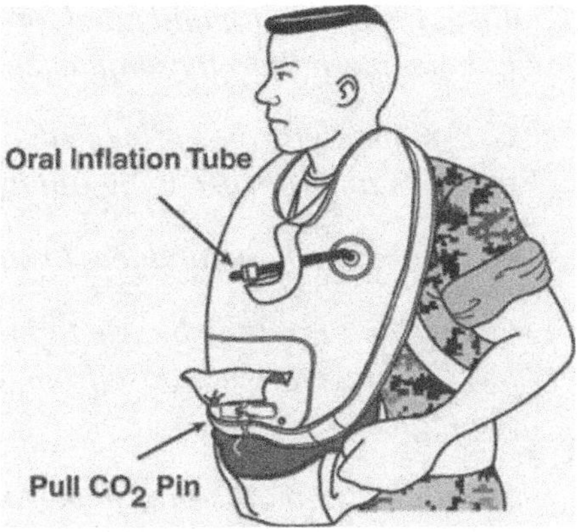

Staying Afloat Without a Life Preserver

You may be in open water without any floating objects or a life preserver to help you survive. If so, uniform blouses and trousers can be made into expedient flotation devices.

Floating With an Inflated Blouse

It is possible to float by a bubble of air trapped in the shoulders of your blouse. The air rises to the back and shoulders of the blouse and supports you at the water's surface. An inflated blouse is also a temporary flotation device used by weaker swimmers while trying to remove their trousers. There is a primary and an alternate way to create a bubble of trapped air in a blouse—

Primary Method
Turn the collar inside the blouse to help create a seal.

Unbutton top button and pull collar around mouth and nose.

Take a deep breath and bend forward slightly at the waist. Exhale one-half to three-quarters of a breath into the blouse.

Grasp and twist the collar with one hand to create a seal, this prevents air from escaping out from the collar.

Use your free hand and feet to stroke and kick to the surface.

Gather and hold the blouse tightly at the collar and stomach level to prevent the blouse from losing air if it floats up too high.

Splash water on the blouse periodically to prevent the material from drying, dry material allows air to escape.

Repeat inflation as required.

Alternate Method
Turn the collar inside the blouse to help create a seal.

Unbutton the second button from the top.

Take a deep breath and bend forward slightly at the waist.

Place your mouth and nose inside the hole created by the open button and exhale one-half to three-quarters of a breath into the blouse.

Grasp material at the unbutton portion and pull downward.

Use your free hand and feet to stroke and kick to the surface.

Splash water on the blouse periodically to prevent the material from drying, dry material allows air to escape.

Repeat inflation as required.

Floating With Inflated Trousers

In warm water, trousers can be used as a primary expedient flotation device. However, in cold water, submerging your head to remove and inflate your trousers results in heat and energy losses that negate the benefit of using the trousers as a flotation device. Once your trousers are inflated, you float motionlessly as if wearing a life preserver. If needed, assume the heat escape lessening posture (known as the HELP position, see page 1-37) to slow heat loss. As trousers dry, air leaks out of the legs. To slow this process, occasionally splash water on the fabric. Reinflate trousers as needed.

Sling Method

The sling method works if you are a strong swimmer or naturally very buoyant. Take the following steps to inflate trousers using the sling method:

Take a deep breath, bend over, and remove your boots.

NOTE: Retain your boots. Tie the boot laces together and suspend the boots from your blouse or hang them around your neck so that they rest on your chest.

Remove your trousers. Button or zip the trouser's fly closed. This allows you to control airflow.

Tie the bottoms of the trouser legs in a square knot. (The appendix illustrates various knots.)

Ensure that the front (fly) of the trousers faces you.

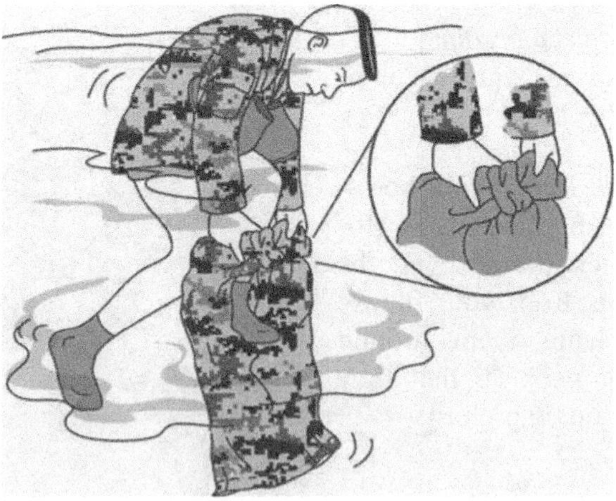

Hold the trousers above the water's surface and behind your head. Grasp both sides of the waistband and open with both hands.

Kick strongly to stay on top of the water while slinging the trousers overhead in order to trap air into them.

Once the waistband is submerged in the water, air is trapped in the legs.

Hold and seal the waistband underwater.

Slip the inflated legs over your head. Hold the waistband in toward your chest, the fly facing your body. To prevent air from escaping from the trousers, seal the waistband by either folding it or twisting it.

Lie back and relax, resting the back of your neck against the knot.

Splash water on the trousers periodically to prevent the material from drying. Dry material allows air to escape.

To replenish air in the trousers, you will use a technique known as the scooping method. With one hand on the open waistband, extend the trousers in front of you just below the surface of the water and scoop air bubbles with your free hand into the open waistband until the trousers have sufficient air. Repeat as necessary.

Splash Method

The splash method is an alternative to the sling method. As with the sling method, you must kick strongly to remain at the surface. To inflate trousers using the splash method, perform the following:

Take a deep breath, bend over, and remove your boots.

Note: Retain your boots. Tie the boot laces together and suspend the boots from your blouse or hang them around your neck so that they rest on your chest.

Remove your trousers. Button or zip the trouser fly closed. This allows you to control airflow.

Tie the bottoms of the trouser legs in a square knot. (The appendix illustrates various knots.)

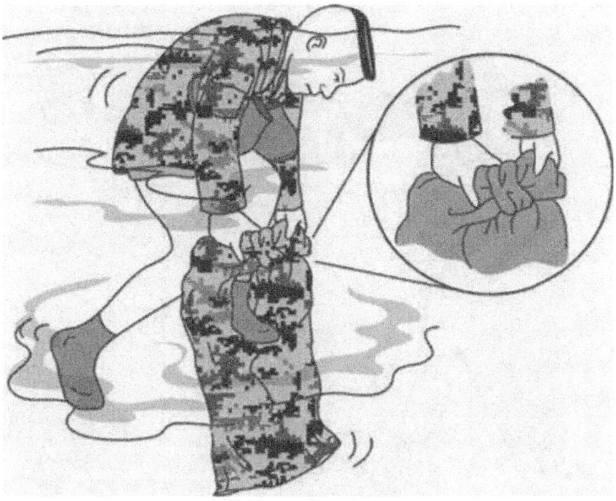

Ensure that the front (fly) of the trousers faces you.

Hold the trousers at the water's surface out in front of you by the waistband with the fly up.

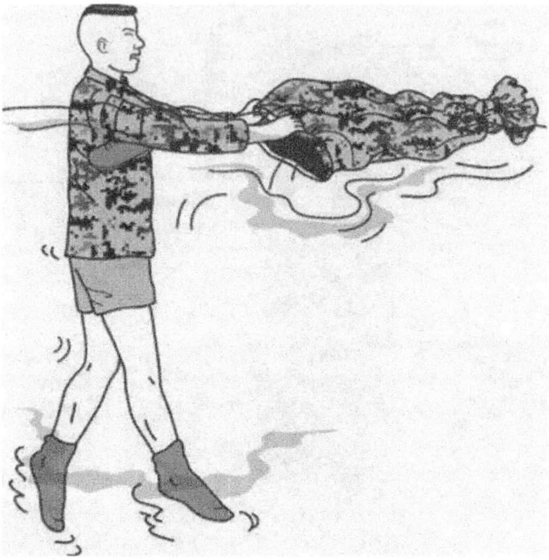

Grasp the waistband at the surface with one hand. Insert your free hand into the waistband, palm down.

Flutter your hand rapidly to create bubbles. This sends a mixture of water and air bubbles into the trousers. The water passes through the fabric. The air remains trapped in the legs.

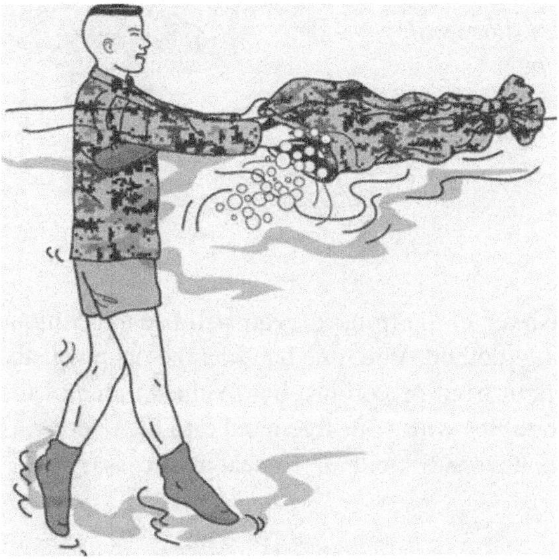

Hold and seal the waistband underwater.

Slip the inflated legs over your head. Hold the waistband in toward your chest, the fly facing your body. To prevent air from escaping from the trousers, seal the waistband by either folding it or twisting it.

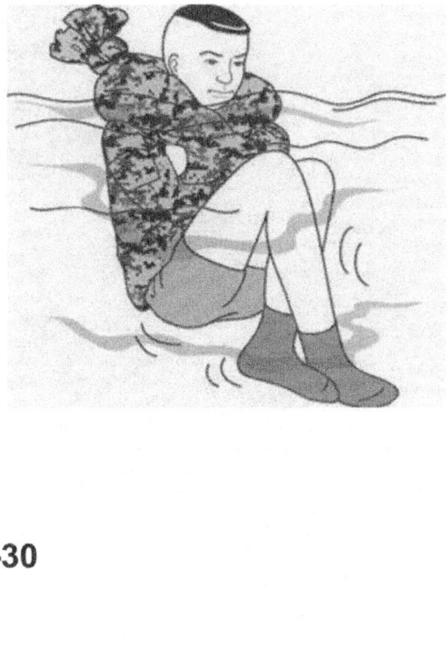

Lie back and relax, resting the back of your neck against the knot.

Splash water on the trousers periodically to prevent the material from drying. Dry material allows air to escape.

To replenish air in the trousers, you will use a technique known as the scooping method. With one hand on the open waistband, extend the trousers in front of you just below the surface of the water and scoop air bubbles with your free hand into the open waistband until the trousers have sufficient air. Repeat as necessary.

Blow Method

The blow method is an alternative to the sling method. Use the blow method if you are a weak swimmer. Take the following steps to inflate trousers using the blow method:

Take a deep breath, bend over, and remove your boots.

NOTE: Retain your boots. Tie the boot laces together and suspend the boots from your blouse or hang them around your neck so that they rest on your chest.

Remove your trousers. Button or zip the trouser fly closed. This allows you to control airflow.

Tie the bottoms of the trouser legs in a square knot. (The appendix illustrates various knots.)

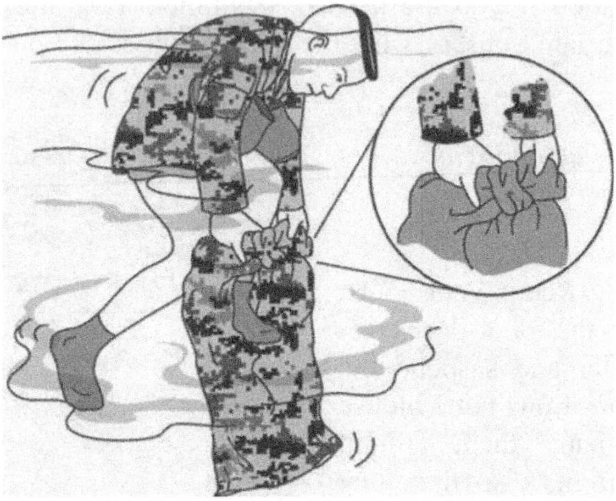

Ensure that the front (fly) of the trousers faces you.

Hold the trousers at the water's surface. Grasp both sides of the waistband and open with both hands.

Take a deep breath.

Drop 2 feet below the water's surface, pulling the waistband underwater.

Hold the waistband open with both hands and blow air into the trousers.

To fill the trousers with air, surface while keeping the waistband underwater, breathe in again, drop below the water's surface, and blow air into the trousers. Repeat these steps until the trousers are filled sufficiently. Once trousers are filled—

Hold the waistband underwater. Twist and pinch it off.

Slip the inflated legs over your head. Hold the waistband in toward your chest, the fly facing your body. To prevent air from escaping from the trousers, seal the waistband by either folding it or twisting it.

Lie back and relax, resting the back of your neck against the knot.

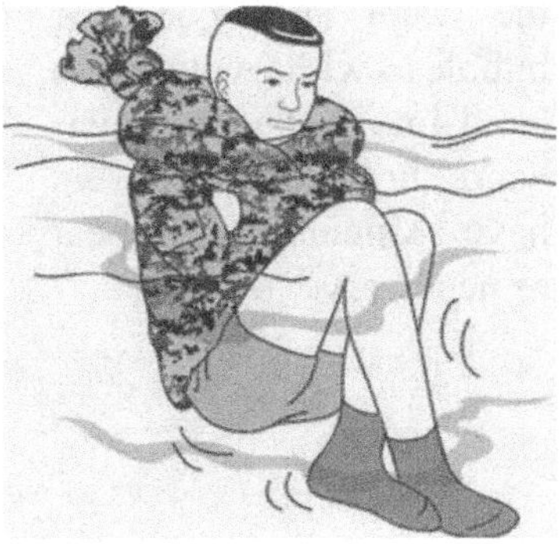

Splash water on the trousers periodically, to prevent the material from drying. Dry material allows air to escape.

To replenish air in the trousers, you will use a technique referred to as the scooping method. With one hand on the open waistband, extend the trousers in front of you just below the surface of the water and scoop air bubbles with your free hand into the open waist-band until the trousers have sufficient air. Repeat as necessary.

Avoiding Heat Loss in Cold Water

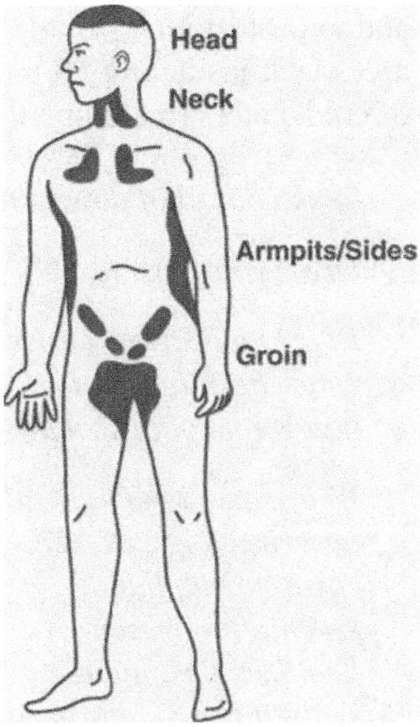

The rate of heat exchange in the water is about 25 times greater than it is in air of the same temperature. When you are immersed in cold water, hypothermia occurs rapidly due to the decreased insulating quality of wet clothing and as a result of water displacing the layer of still air that normally surrounds the body. You also lose about 50 percent of your body heat through your head; therefore, keep your head out of the water. Other areas of high heat loss are the neck, the armpits/sides, and the groin.

In cold water, DO NOT SWIM TO STAY WARM. Swimming, even with a slow and steady stroke, produces a lot of heat that is lost in the water. The heat loss can produce hypothermia that slows body functions and can result in serious injury or death. Remaining motionless conserves body heat three times longer than swimming. SWIM only if you have a flotation device and the shoreline is visible.

Individual Protection From the Cold

If you are equipped with a life preserver, assume the heat escape lessening posture (known as the HELP position) to slow heat loss and to protect major blood vessels near the body's surface. These areas lack insulating fat and are vulnerable to the chilling effects of cold water. To assume the HELP position—

Tuck your chin down tightly to cover your throat.

Draw your legs up in a fetal position to protect the groin.

Place your arms across your chest, tuck your hands into your armpits.

Wear some type of head covering (e.g., stowed cover, towel, handkerchief) to lessen heat loss through the scalp if head covering is available.

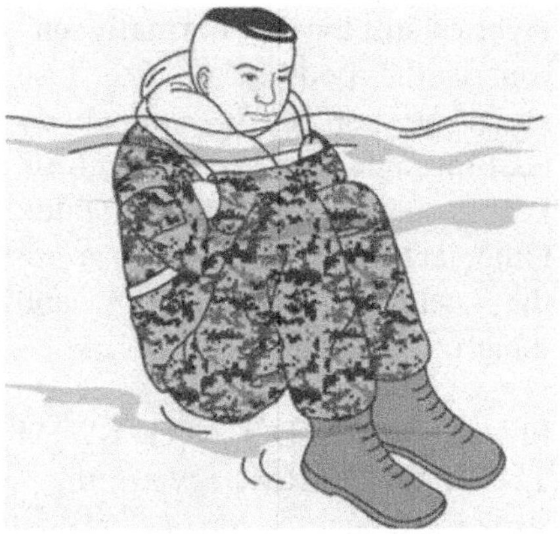

Group Protection From the Cold

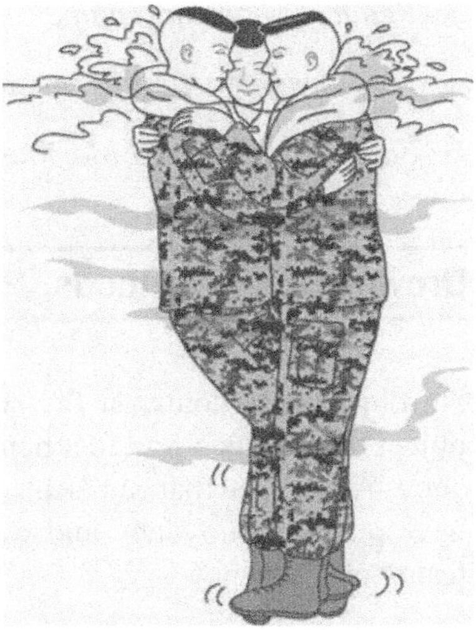

If three or more Marines are in the water and are equipped with life preservers, they should wedge tightly together and lock arms to form a circle known as a huddle position. This position protects vulnerable areas from heat loss.

A casualty who is suffering from the effects of the cold can be placed within the huddle to be surrounded by warmer water. If in the water for a prolonged period, it is recommended that everyone be rotated inside the huddle to maintain or re-warm each person's internal core temperature. If there are more than five Marines, they should make clusters of huddle positions.

Contact with other swimmers provides survival advantages—

Creates a larger target for search and rescue aircraft.

Provides additional warmth in cold water.

Improves morale.

Re-establishes the chain of command.

Reduces shock and panic.

Provides opportunities to administer first aid.

Supports exhausted Marines.

Drownproofing Methods

An object that floats on the surface has positive buoyancy. An object that floats a few feet beneath the surface has neutral buoyancy. An object that sinks has negative buoyancy. Most people have positive buoyancy and will float at the water's surface. To test your buoyancy—

Stand in water that is at your chest level or deeper.

Take a full breath.

Bend forward slowly.

Relax and wait.

If you have positive buoyancy, you will slowly rise to the surface. If you have neutral buoyancy, you will float a few feet beneath the surface. If you have negative buoyancy, you will sink. Regardless of whether you naturally float or sink, you can remain at the surface for extended periods without a life preserver if you exercise the appropriate drownproofing methods or survival strokes, which are based on your buoyancy.

Drownproofing methods consist of the T-method and the sweep. The breast stroke, side stroke, and elementary backstroke are the most common survival strokes. With any drownproofing method or survival stroke, remember the acronym **SAFE**:

SLOW EASY MOVEMENTS
Move slowly to conserve energy and minimize heat loss.

APPLY NATURAL BUOYANCY
Use natural buoyancy to support the body.

FULL LUNG INFLATION
Fill the lungs with each breath. Do not hold air in the cheeks.

EXTREME RELAXATION
Tight muscles are denser than relaxed ones and do not float as well.

Crawl Stroke

The crawl stroke, sometimes called the front crawl or free style, is the fastest stroke. To execute the crawl stroke—

Body position
Lie horizontal, on your stomach, in the water.

Look forward and downward at a 45 degree angle with the waterline between your eyebrows and hairline. Your head position is important as it assists in cutting a path through the

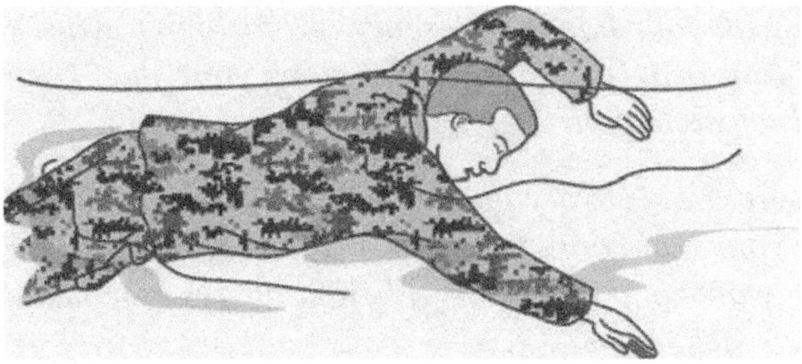

water. If your head is too high in the water, your lower body sinks significantly, making your stroke less efficient. If your head is too low in the water, water washes over your shoulders and neck, causing unnecessary drag.

Arm action

Arm action occurs in three phases: catch, propulsion, and recovery.

Fully extend one arm forward of your body, this positions your hand to catch the water in preparation for the propulsion phase.

To catch the water, bend your wrist (with your palm pointing outboard) and make an "S" shape (or inverted "S" shape) with your hand, ensuring that your hand does not cross the center of your body. Your left hand makes an "S" shape and your right hand makes an inverted "S" shape with your hand finishing at shoulder level.

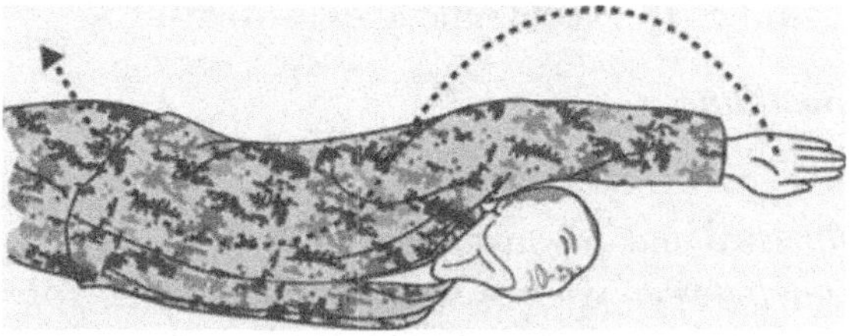

Push with your hands in a rearward fashion toward your feet until your arm is fully extended along your side, keeping your hands close to your body.

To begin the recovery phase, bend your arm at the elbow and raise your hand out of the water. Your hand breaks the surface of the water and maintains a height of 2 to 3 inches above the water's surface.

With your hand and arm moving in a forward manner, bring your hand past your head until your arm is about three-quarters of the way extended. At this point, turn your hand with your palm outboard so as to allow your thumb and fore-finger to enter the water first.

Once your hand enters the water, continue to push your arm forward until it is fully extended.

You are now prepared to catch the water again. These steps are performed in an alternating pattern: when one arm is catching and propelling, the other arm is recovering.

Leg action
Use a flutter kick to create the leg action for the crawl stroke. This kick is used for both propulsion and keeping the lower body horizontal with the water's surface. The flutter kick is an alternating leg action: one leg is kicking in a downward motion while the other leg is recovering to the surface to prepare for the next kick. The size of the kick ranges from 12 to 15 inches and depends on your height.

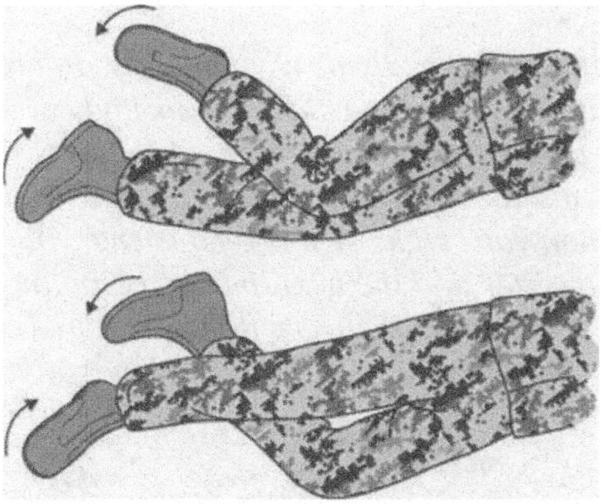

Maintain your legs in a semi-rigid manner.

Generate power for the kick from the hips.

Keep your feet loose at the ankles, they trail behind your legs and act as "flippers."

Execute the propulsion phase of this kick with a downward thrust of your leg.

Execute the recovery phase by pushing your leg back to the surface. This phase is complete when your foot reaches the surface.

Breathing
Breathe during the recovery phase of the arm action.

Roll your body and rotate your head to the side of your body where the arm recovery is occurring, this rolls the water away from your mouth. Keep your chin pushed back toward your shoulder.

Exhale while your face is still submerged and inhale when your face breaks the surface.

Breathe either bilaterally or rhythmically. To breathe rhythmically, breathe on the same side of your body every time your arm cycle occurs. Bilateral breathing requires you to breathe every one and a half arm cycles. To breathe bilaterally, breathe when your right arm is recovering, your face goes back in the water and the next time you breathe is when your left arm is starting its recovery three arm strokes later.

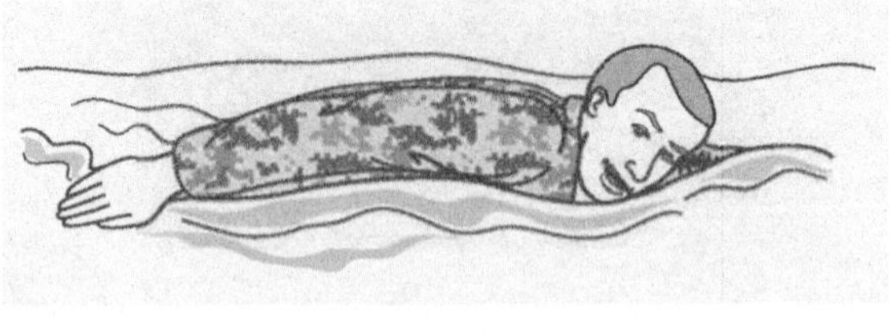

Bilateral breathing is the preferred method for breathing; it prevents the chance of hyperventilation and allows your body to maintain a lateral position in the direction you are swimming.

Coordination

This stroke uses constant arm and leg action.

T-Method

The T-method is a basic drownproofing method. This is the best survival technique if you have negative buoyancy. To execute the T-method—

With your face out of the water, take a deep breath and submerge your face in the water while holding your breath.

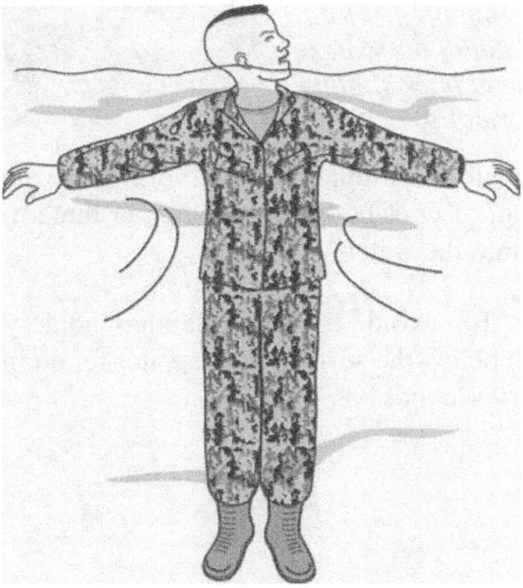

Float with your body in a horizontal position, arms extended from your side, and legs extended and joined.

Move your hands up to your armpits by tracing an imaginary line along your ribs.

Extend your arms outward (horizontal) to your sides, your body position resembles the letter "T."

Step out and forward with one leg and point your other leg to the rear, your knees should be slightly bent. Simultaneously, bring your arms down to your sides. Then bring your legs back together. You complete the steps by exhaling most of your air and preparing to surface your face to obtain another breath of air. Hold your head out of the water. Tilt your head back slightly. Breathe normally.

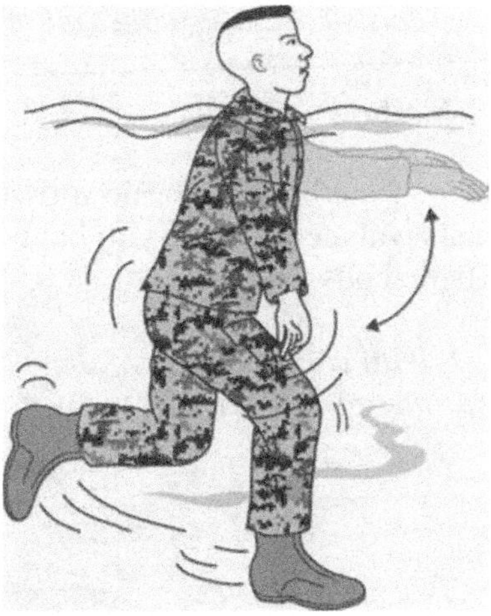

Once your breath is complete, move your hands up and down directly in front of your body. Do this two or three times to slow your descent into the water.

> NOTE: To avoid hyperventilating, hold your breath below the surface of the water for no more than 10 seconds.

The Sweep

The sweep works well if you have slight to excellent positive buoyancy. To execute the sweep—

Float face down in the water: bend 45 degrees at the waist, arms and legs dangling, head hanging down, relax all muscles.

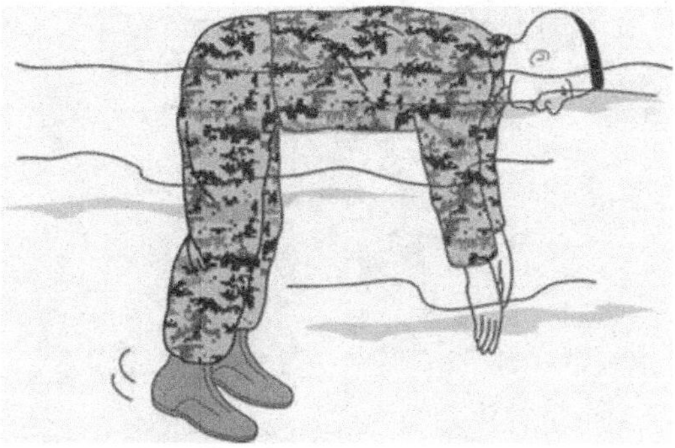

Spread your feet slowly to prepare for a single kick (one leg is forward and one leg is rearward).

Cross your arms in front of your chest, palms outboard with the back of each hand touching the opposite ear.

Exhale prior to raising your head for a breath.

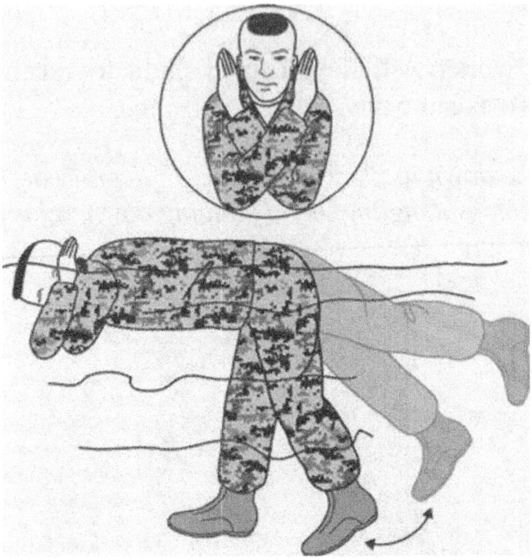

Bring your legs together and sweep your arms down and out until the arms are fully extended out to the sides. This raises your face above the water and allows you to catch a breath of air.

With your air supply replenished, return your face to the water and relax while sweeping your arms in a downward motion in front of your body to prevent/slow your descent. Do not hold your breath for more than 10 seconds. More than 10 seconds enhances your chances of shallow water black out and subsequent drowning.

Breast Stroke

Use this stroke to swim underwater, through oil or debris, and in rough seas. If you are a good swimmer and not wearing combat gear, the breast stroke is the best stroke for long-range swimming because it provides good visibility and allows you to conserve your energy and maintain a reasonable speed.

Body position
Lie prone in the water. Swim with your trunk and legs projecting back and down at an angle of 20 to 30 degrees.

Extend arms out in front (hands together [side by side]), and extend legs behind (toes pointed) to prevent drag.

Face downward, looking forward at a 45 degree angle to break the water and to prevent water from washing into the collar area causing drag.

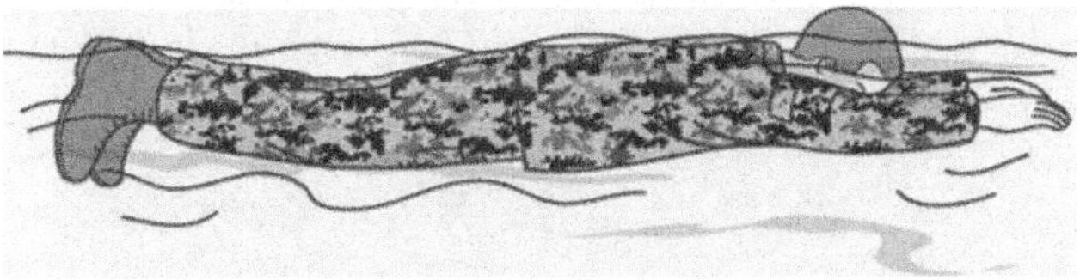

This is known as a glide.

Arm action

Turn your palms outward and bend your arms slightly.

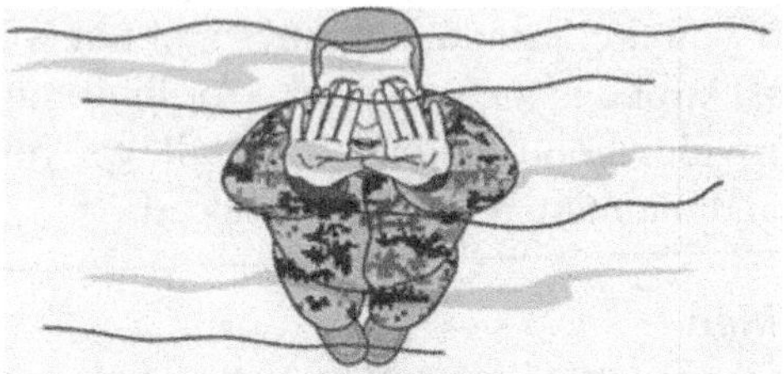

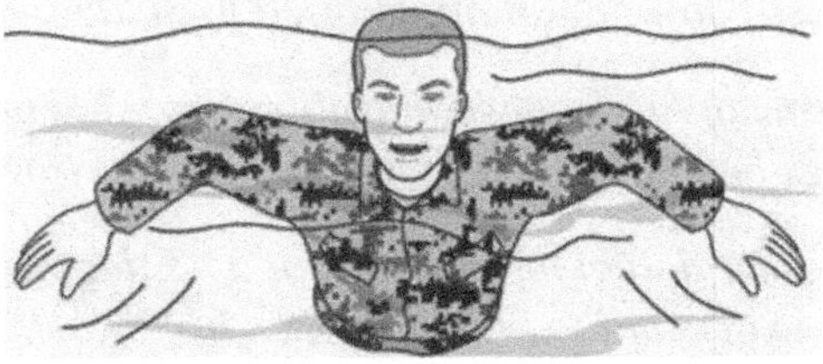

Sweep your arms sideward and slightly downward until your hands are opposite and slightly below your shoulders.

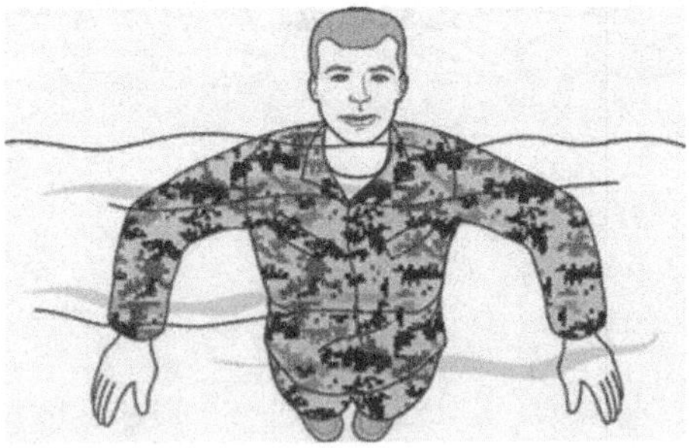

Rotate your head up, breathe once your mouth breaks the surface.

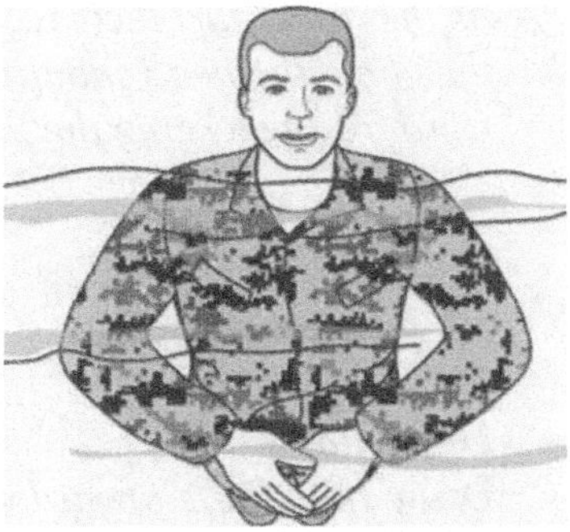

Bring your hands and arms up along your chest and thrust them forward until they are extended and ready to execute the next arm pull.

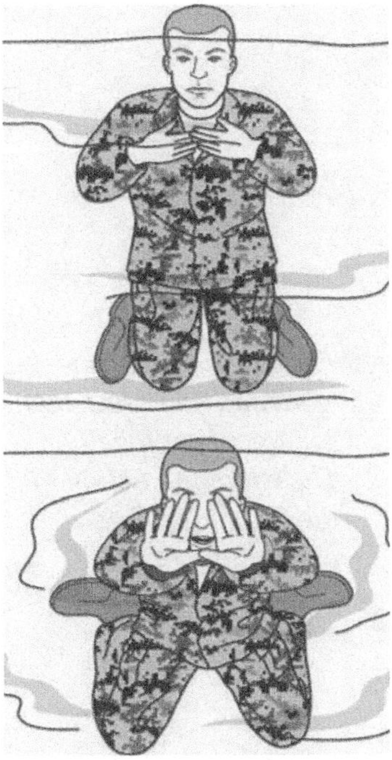

As the arms start their recovery into the glide, the head should rotate forward, resubmerging the face.

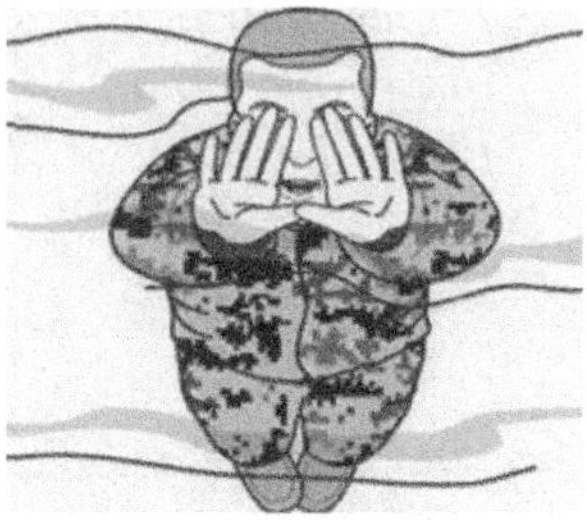

Leg action

Draw your heels toward your buttocks, establish a 45 degree bend in the knees.

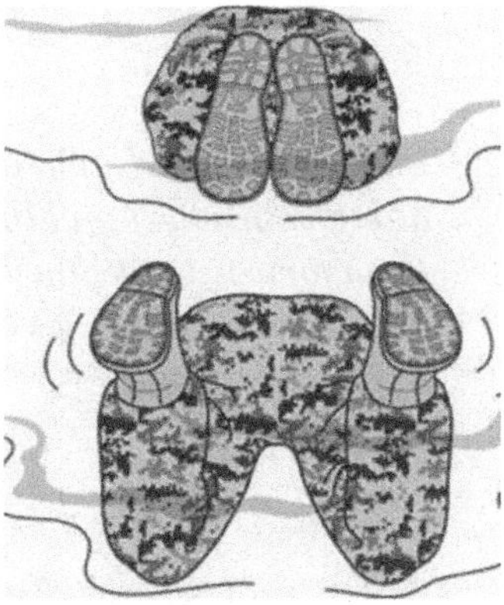

Thrust your legs outward and rearward, then squeeze them together. The whipping action of the feet aids forward propulsion.

This is known as the breast stroke kick.

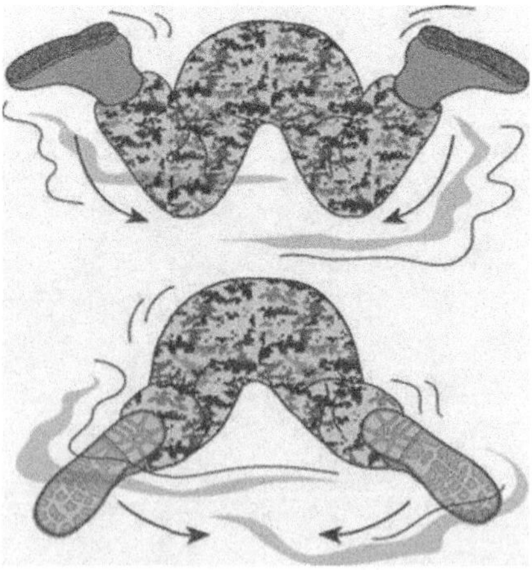

Breathing
Inhale during the arm pull and exhale through your mouth and nose during the finish of the breast stroke kick and glide.

Coordination
The stroke movement is in three counts:

Begin your arm pull. Near the finish of the pull, flex your knees and bring your heels toward your buttocks. The arm pull counteracts the resistance created by the knees.

As the arm pull is completed, thrust your hands forward, kick your legs outward and rearward, and squeeze them together.

Glide through the water for approximately 1 to 3 seconds or until your forward momentum decreases, then begin the next stroke.

Side Stroke

The side stroke is a survival stroke because you use both arms for buoyancy, with each arm creating a slight propulsion. The majority of your body's propulsion comes from your kick. To execute the side stroke—

Body position
Lie on your side with your lead (bottom) arm extended beyond (with a slight bend in your elbow) your head and in line with your body. Palm is down and your hand is submerged 6 to 8 inches.

Extend your trail (top) arm down the length of your body over your thigh.

Keep your legs straight and together, toes pointed rearward.

Keep your face out of the water, this allows for free breathing.

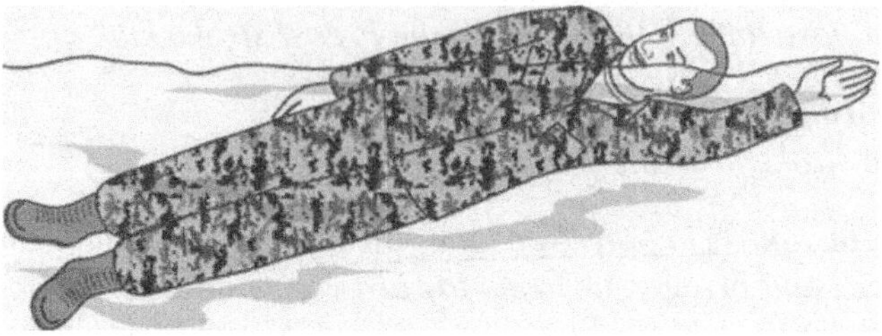

This is known as the glide.

Arm action

With your lead arm, pull your arm downward, while flexing at the elbow, until it is straight down from your shoulder.

Rotate your shoulder and pull your elbow into your side. This should put your lead hand at shoulder level. At the same time, turn your palm toward your face and thrust forward to your original, extended position.

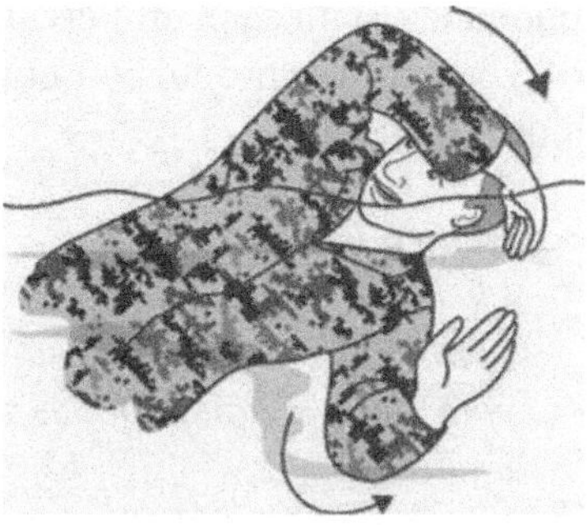

Draw your right hand upward in front of your chest to shoulder level. Rotate your palm toward your feet, then push it downward in front of your body toward your feet to catch the water.

Push your trail hand backward to its original position on top of your thigh. (Your trail hand starts forward and meets your lead hand at your chest/shoulder.)

Leg action

To perform the scissor kick, the top leg always goes forward and the bottom leg always goes rearward. From the extended position, draw—or recover—your feet toward your buttocks until your legs are bent at a 45 degree angle at the knees and the hips are flexed at a 45 degree angle with the thighs.

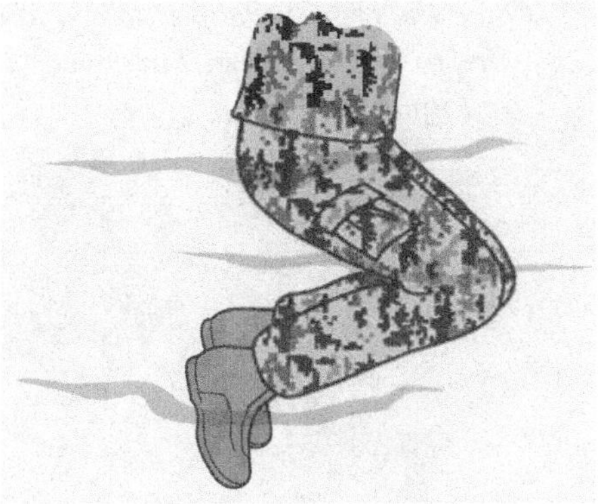

Once the legs have completed their recovery and while maintaining a 45 degree bend in the knees, extend the legs fully into a "V" shape in order to catch the water for the propulsion phase.

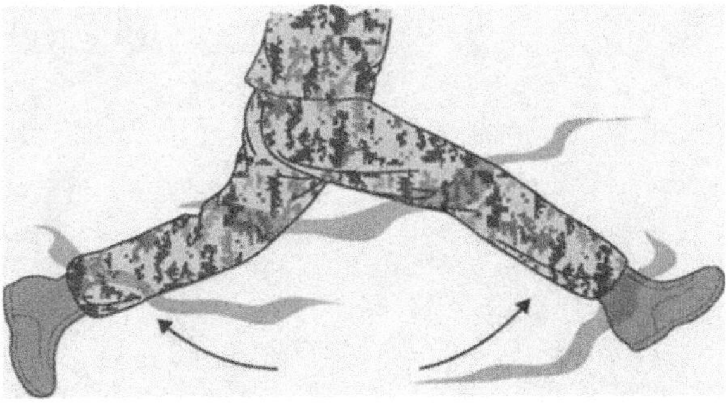

Once the legs are separated and extended forward and rearward to the "V" position, sweep the legs together until the feet are together.

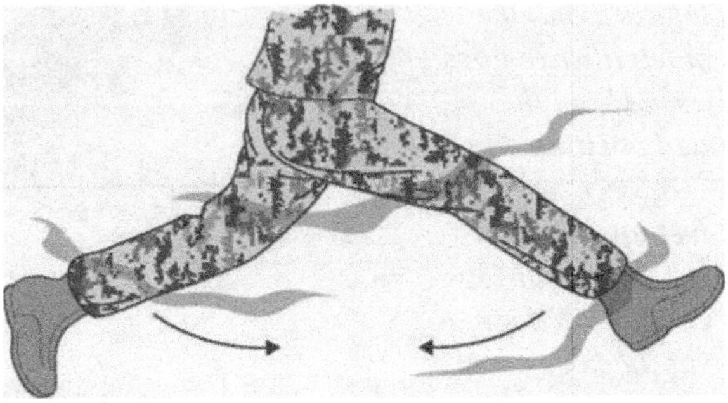

You are now in the glide position.

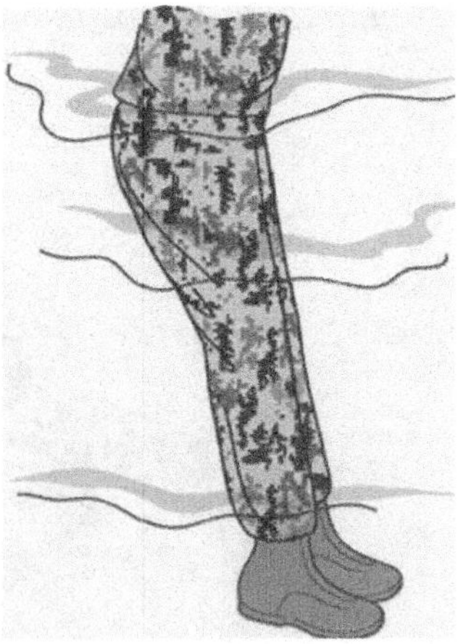

Breathing

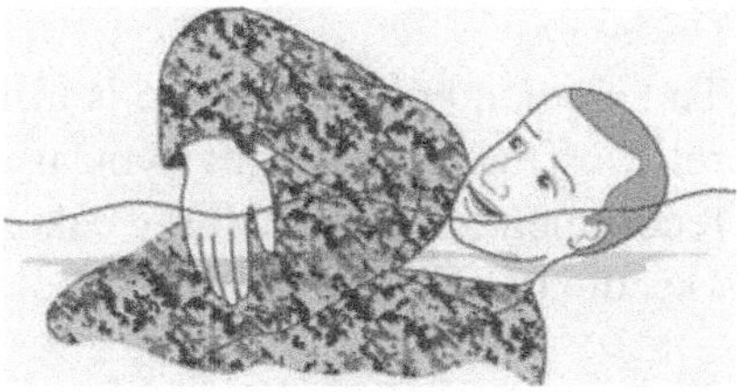

As long as your face remains clear of the water, it is a free breathing stroke. However, it is recommended that you exhale then quickly inhale when the legs are sweeping back together in the scissor kick. This is when the body reaches its highest point in the water, thus clearing the face completely from the surface of the water making it the optimum time to breathe.

Coordination

Begin the stroke with the downward pull of your lead arm. At the same time, bring your trail arm upward and draw your knees up to begin the kick. Let the thrust of the lead arm, push of your trail arm, and the kick of your legs coincide in order to finish the glide position. Glide through the water for approximately 1 to 3 seconds or until your forward momentum decreases, then begin the next stroke.

Elementary Backstroke

The elementary backstroke is also an excellent survival stroke. It relieves the muscles that you use for other strokes, and it is the recommended stroke for weak swimmers or nonswimmers. To execute the elementary backstroke—

Body position
Start on your back.

Face up, chest up, and hips up, keeping an arch in your lower back with arms pressed to your sides and your legs extended and joined to prevent drag.

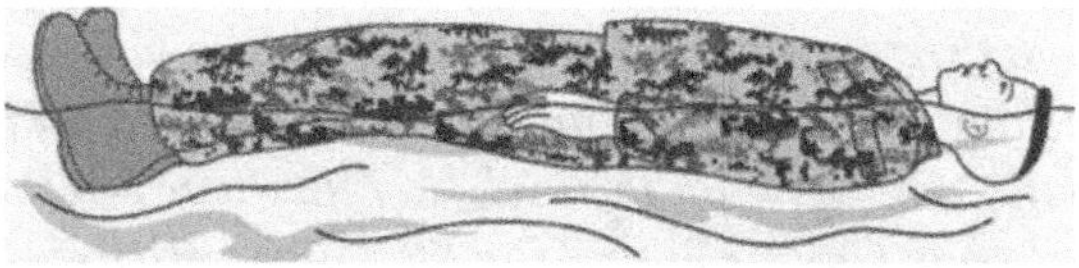

Arm action
Trace your hands up your sides to an area near your armpits then extend your arms out to the sides to form the letter "T" (palms facing feet), locking out the elbows.

> NOTE: Don't raise your arms above your head. This creates drag, changes your body position, and submerges the head.

Slap your palms to your thighs using a strong sweeping motion.

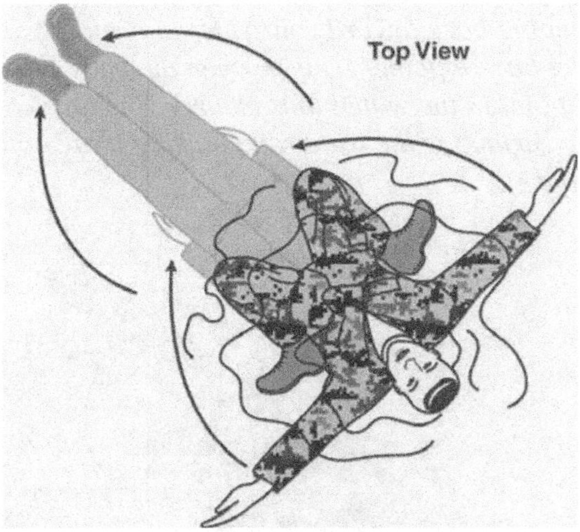

Leg action
Bend both legs at the knee (90 degree angle) slightly separating your knees and drawing your heels downward to a point under and outside your knees. The knees are spread as wide as the hips or slightly wider depending on the body type of the swimmer.

Circle around in a whipping action, ending with legs in a glide position.

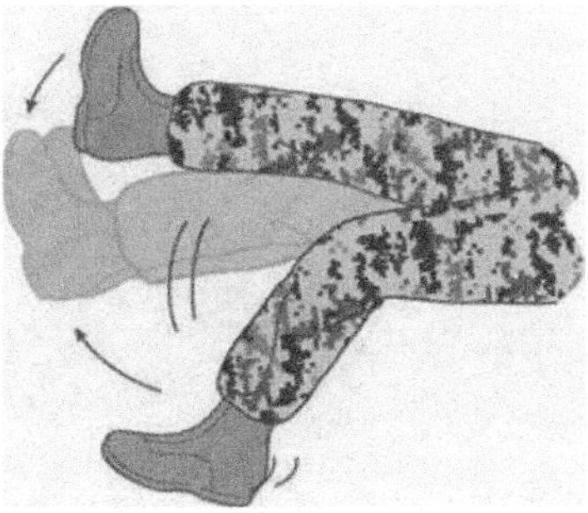

Breathing

Breathe anytime during this stroke. However, it is recommended that you exhale then quickly inhale when your arms are sweeping back toward your sides and while your legs are sweeping back together. This is when the body reaches its highest point in the water, thus clearing the face completely from the surface of the water and making it the optimum time to breathe.

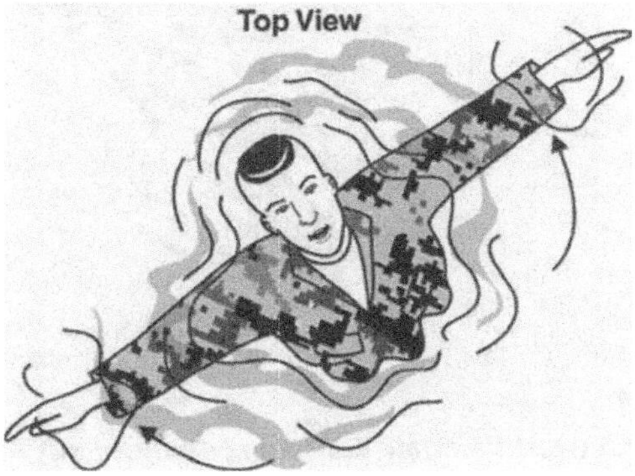

Coordination

The stroke movement occurs in three counts (recovery, catch, power).

Begin the arm pull (recovery).

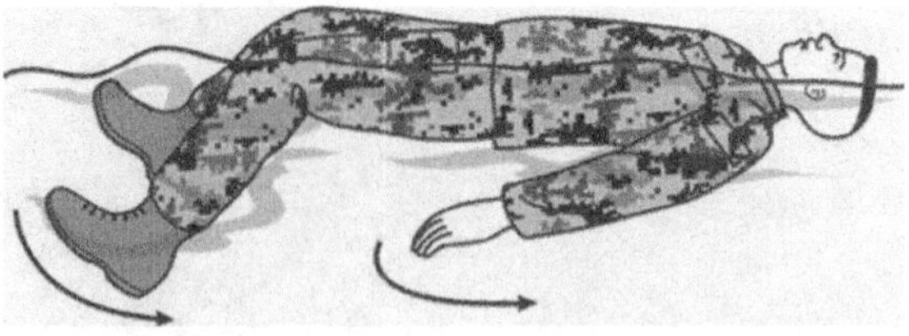

Near the finish of the pull, flex your knees to a 90 degree angle. The arm pull counteracts the resistance created by the knees.

Kick out your legs, and squeeze them together as the arm pull is completed (catch, power).

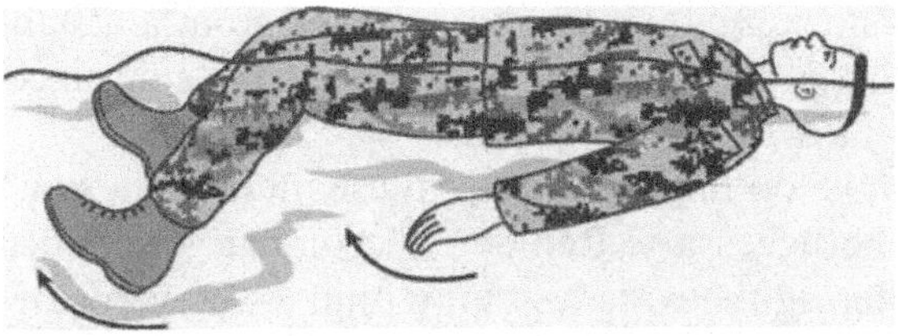

Glide through the water for 1 to 3 seconds or until your forward momentum decreases, then begin the next stroke as your momentum slows.

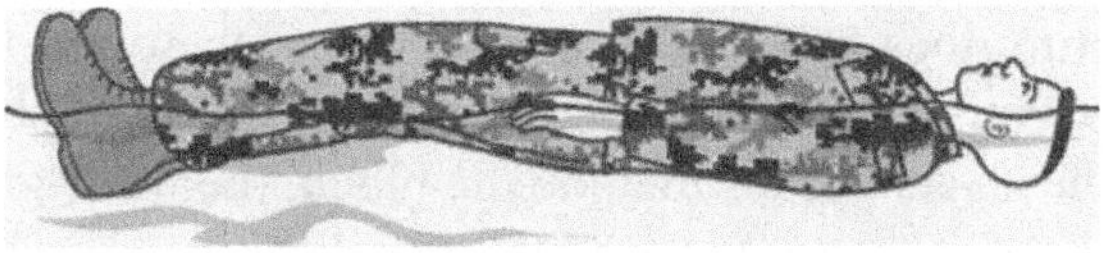

Chapter 2
Water Rescues

A drowning victim can panic and react with unexpected violence and can seize and inadvertently drown a rescuer. Therefore, if possible, a water rescue should be executed from a distance. Reaching, wading, or throwing methods are used in a distance rescue.

If a victim is too far away to use these methods, a swimming rescue may be necessary. Remove all combat gear, when possible, before entering the water. Swim within 2 to 6 yards of the victim to maintain a margin of safety, this allows you to reassess the situation and reassure the victim. If the victim is unconscious, use the wrist tow method or cross-chest carry method to pull the person to safety. If the victim is struggling, use a rear approach and then execute either a single armpit level off or a double armpit level off before towing the victim to safety.

If the victim does begin to overpower you, there are techniques that allow you to defend yourself without having to abandon the rescue. These techniques include the block, the wrist-grip escape, the front head-hold escape, and the rear head-hold escape. These techniques allow you to separate yourself from the victim, reassess the situation, and then attempt the rescue again.

> *Note:* In this chapter's illustrations, the rescuer is shown without a helmet.

Reaching Rescue Techniques

Reach

From a safe position at the water's edge, reach out to the victim. Talk constantly to calm the victim. Retain partial contact with land or some solid support structure (e.g., pier, bridge). If the victim is close but still beyond reach, extend an object (e.g., stick; pack; rifle with magazine removed, chamber empty, and muzzle pointing toward victim) that the victim can grasp. Pull the victim slowly to safety. Once the victim is close to shore, remove the victim from the water.

NOTE: You can also extend a foot to the victim if you can retain a secure grip on a solid support structure.

Reach From a Deck

Reach from a deck rescue can be executed by either a swimmer or a nonswimmer, and it can be used on an active or a passive victim. Be sure to reassure the victim during the rescue. To execute reach from a deck—

Lie prone on the deck with your body firmly anchored. To anchor your body, lie flat, spread legs apart, and extend one arm behind you with your palm down and on the deck.

Keep as much of your weight on the deck as possible and extend your free hand to the victim.

Grasp the victim's wrist from above, your thumb and index finger are facing you. The victim should never be allowed to grab you and put your life at risk. Therefore, never reach across the victim to grab his wrist, always grab the wrist that is the closest to you.

Keep your arm straight and locked out and pull the victim to the side of the deck. The victim should never be allowed to grab you and put your life at risk. Therefore, never pull the victim into you, always pull the victim into the side.

Arm Extension

An arm extension rescue can be used if you cannot reach the victim using the reach from a deck rescue technique and you must enter the water. An arm extension rescue technique is used for a victim who is either active or passive. Once you determine that a reach from a deck rescue technique is not viable—

Reassure the victim and quickly ease into the water while holding onto the deck with one hand.

Grasp the victim's wrist from above, your thumb and index finger are facing you. The victim should never be allowed to grab you and put your life at risk. Therefore, never reach across the victim to grab his wrist, always grab the wrist that is the closest to you.

Keep your arm straight and locked out and pull the victim to your side. The victim should never be allowed to grab you and put your life at risk. Therefore, never pull the victim into you, always pull the victim into the side.

Leg Extension

If the victim is beyond the reach of your arm, ensure that you have a firm grip on the deck, extend a leg to the victim and allow him to grab it, slowly bring the victim in closer until you can grab the victim's wrist that is holding onto your leg. Once you have a firm grasp on the victim's wrist, use the steps in the arm extension to pull the victim to safety. The leg extension rescue can only be used to rescue an active victim because the victim must be able to grab the rescuer's leg.

Wading Assist

Do not wade into water that is deeper than your chest. Talk constantly to calm the victim. If possible, do not touch the victim directly. Extend an object (e.g., stick; pack; rifle with magazine removed, chamber empty, and muzzle pointing toward victim) that the victim can grasp. Once the victim grasps the object, pull the victim slowly to safety.

Throw

If the victim is not within reach, use an expedient line to throw a lifesaving device to the victim. A lifesaving device can be any weighted item that floats (e.g., a canteen that is one fourth full of water). The lifesaving device must be secured to the end of the rope so that the rope will feed out from its coil when tossed. Talk constantly to calm the victim. Once the victim grasps the line or the lifesaving device, pull the victim toward you at a steady pace that keeps the victim's head above the water's surface. DO NOT pull so strongly that you break the victim's grip on the line. To prepare and use an expedient line and lifesaving device—

Tie a bowline at one end of the rope. (The appendix illustrates various types of knots.)

Unfasten the lid of a canteen.

Place the bowline around the neck of the canteen.

Refasten the lid so the canteen hangs from the bowline loop.

Place one end of the rope under the ball of your forward foot to secure it, either tie a knot at the end of the rope or tie an object at the end of the rope to create a block. Stand with your weight on the end of the rope. DO NOT tie the rope around your ankle.

Coil 20 to 30 yards of the rope, and hold it in a nonthrowing hand.

Place the canteen in your throwing hand.

Use an underhand throw to pitch the canteen and rope a short distance over the victim's head. Keep your nonthrowing hand open so the coil can unfold freely. The rope should trail across the victim's outstretched arms.

Retrieve the rope if the throw is inaccurate or the victim fails to grasp it. Recoil the rope as it is retrieved.

Divide the coil and throw again.

Lifesaving Approaches

Properly approaching the victim is as important as any other aspect of a rescue. Knowing the appropriate and safest method needed to approach a victim is critical in ensuring your safety and the victim's survival. Determining the victim's physical state (distressed swimmer, victim that is active and drowning, or victim that is passive and drowning) is crucial and will determine which type of approach you will execute.

Front Surface Approach

The front surface approach is typically performed when the victim is passive. You must remember that approaching the victim from the front is dangerous because a distressed swimmer or victim that is active can lunge toward you.

Rear Approach

If the victim is active, you must remember that his response can change rapidly. Therefore, approaching him from the front could be extremely dangerous. The victim could easily grab you and take you under the water during a state of panic. Therefore, use a rear approach when possible. When approaching a victim from the rear, reassure the victim until you get to within 2 to 6 yards of the victim; at that point, stop talking to the victim in order to maintain an element of surprise.

Approach Strokes

Before approaching a victim, you must evaluate the victim:

- Is the victim a distressed swimmer?
- Is the victim active and drowning?
- Is the victim passive and drowning?

Select your approach stroke based on your evaluation. Remember, the safety of the rescuer is paramount. Do not endanger yourself in an attempt to reach a drowning victim.

Crawl Stroke Approach Stroke

The crawl stroke (free style) is the fastest approach stroke. This stroke is used when the victim is passive and/or unconscious in the water. That is, the victim is face down, submerged or near the surface of the water, is not breathing, or is not moving. If any of these conditions exist, there is an obvious need to reach the victim as quickly as possible.

Execution of the crawl stroke approach stroke is the same as the crawl stroke (see pages 1-40 through 1-44), except that your head remains above the water's surface to allow for free breathing and maintaining eye contact with the victim.

Maintaining eye contact is critical. If the victim becomes submerged, you will have a better chance of locating him if you know his last location. Your distance from the victim will determine if your head remains out of the water the entire time you are executing the stroke. If you are 54 yards or less from the victim, you should maintain eye contact by keeping your head raised until you reach the victim. If you must swim more than 54 yards to reach the victim, make eye contact with the victim, place your face back in the water for four or five strokes, then raise your head again to regain eye contact with the victim (repeat these steps until you reach the victim).

Body position
The position of your body is horizontal to diagonal and influenced by the position of the head. This allows you to maintain eye contact with the victim.

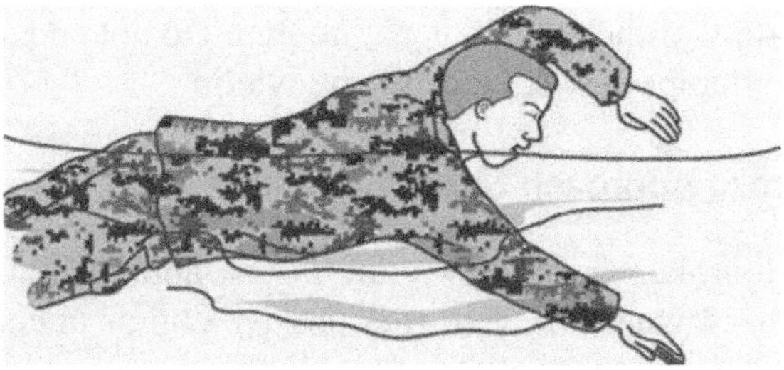

Arm action
While one arm is forward of your head to catch the water for the propulsion phase, extend your other arm alongside your body in what is known as the recovery phase. The recovery phase ends when your arm is extended forward of your head to begin the propulsion phase. In the propulsion phase, your arms are approximately shoulder-width apart and the pulling action is slightly wider and deeper to compensate for your raised head.

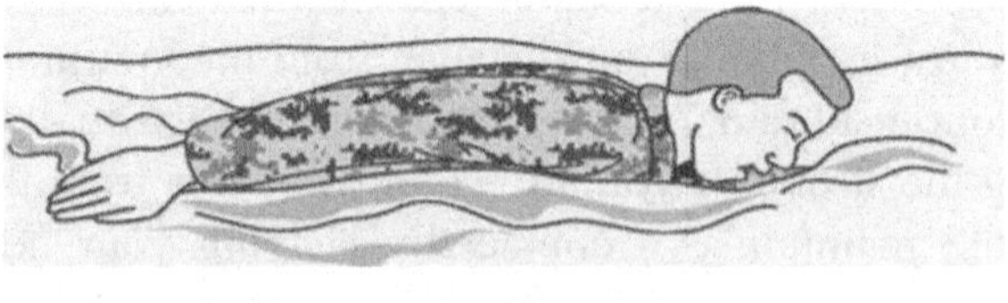

Leg action
Kick one leg in a downward motion while your other leg is recovering to the surface to prepare for the next kick (this is known as a flutter kick). This kick is used for propulsion and maintaining the lower body as horizontal to the water's surface as possible. When your head is raised, your knees are slightly bent in order to keep your feet near the surface.

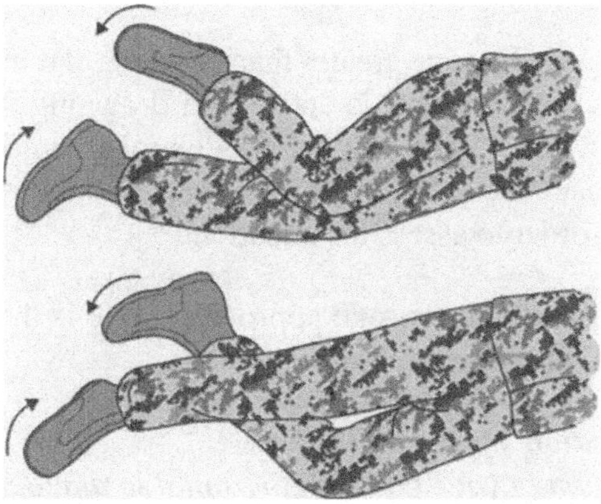

Breathing
If your face is above the water, breathe as needed (free breathing).

If traveling in excess of 54 yards to reach the victim, rotate your head to the right or left during the recovery phase and inhale when your face is out of the water or wait until your head is out of the water and facing forward while regaining sight of the victim.

Coordination
This stroke uses constant arm and leg action.

Breast Stroke Approach Stroke

The breast stroke approach stroke is used when a swimmer is distressed or a victim is active and drowning. This stroke is also used when the victim has a suspected spinal injury because it minimizes the wake created around the victim, minimizes the movement of the victim's head and neck, and helps prevent further injury.

If a victim is at a distance greater than 54 yards and if the swimmer is in distress or the victim is active and drowning in open water (ocean, lake), you should use the crawl stroke approach stroke to rapidly approach the victim, stopping approximately 11 yards from the victim in order to assess the situation.

Execution of the breast stroke approach stroke is the same as the breast stroke with a few minor deviations.

Body position
The position of your body is horizontal to diagonal and influenced by the position of the head. This allows you to maintain eye contact with the victim. It also allows you to communicate with the victim in order reassure him as you approach.

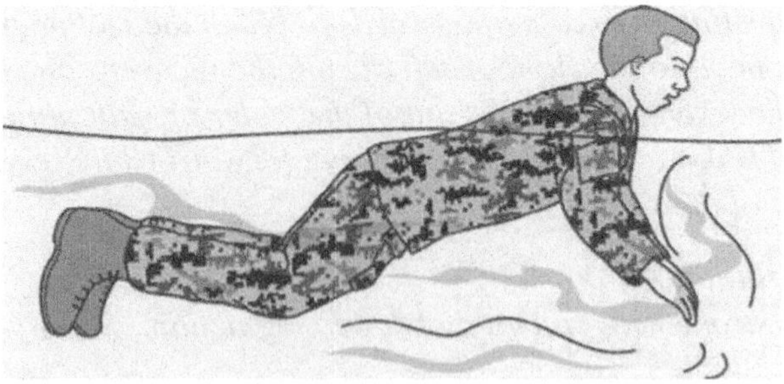

Arm action

Execute the arm action for the breast stroke, except that your arm pull is wider and deeper, which allows your head to remain above the water's surface.

Leg action
Execute the leg action for the breast stroke, except that your legs open wider during the propulsion phase of the stroke.

Breathing
Keep your face above the water and breathe as needed (free breathing).

Coordination
Coordination is the same as the breast stroke, except that there is no glide period. You constantly stroke until you reach the victim.

Level Offs

A properly performed level off positions the victim's face up and horizontal to the water's surface. There are three types of level offs: front surface approach, single armpit level off, and the double armpit level off. The single armpit level off and the double armpit level off are done by performing a rear approach, which is the preferred approach when a victim is active.

Front Surface Approach

The front surface approach is performed when the victim is passive. Approaching a victim from the front may place you in danger because a distressed swimmer or a victim who is active may lunge toward you. To execute a front surface approach—

Stop 2 to 4 yards from the victim to reassess the situation and reassure the victim.

Determine which arm to use in order to rotate the victim into a face-up position (your right hand on the victim's right wrist or your left hand on the victim's left wrist).

Move into position and prepare to grasp the victim.

Turn sideways and move toward the front of victim.

Once in position, reach forward and grab the victim's wrist. Your thumb is on the inside of the victim's wrist, as if you were checking his pulse with your thumb. Your remaining fingers wrap around the victim's wrist.

Lean back immediately and execute a powerful scissor kick or inverted scissor kick and perform short, vigorous pulls with your free arm.

As the victim begins to move forward in the water, kick, pull, and twist outboard on the victim's wrist. The momentum created from the kick and the pulling and twisting action of the victim's wrist will rotate the victim into a face-up and horizontal position in the water.

Extend and lock out your towing arm down the length of your body and execute a wrist tow to move the victim to safety.

Single Armpit Level Off

Approach the victim.

Stop 2 to 6 yards from the victim to reassess the situation, reassure the victim, and maintain your margin of safety.

Approach the victim slowly and grab his armpit with your hand (your right hand to victim's right armpit or your left hand to victim's left armpit). Position yourself sideways to the victim and place your elbow in the center of his back.

Pull with your hand that is in the victim's armpit while pushing with your elbow that is in the victim's back to place the victim horizontal on the water's surface (face up). To assist in placing the victim horizontally, use your free arm to execute short, vigorous pulls and use your legs to execute a scissor kick or inverted scissor kick.

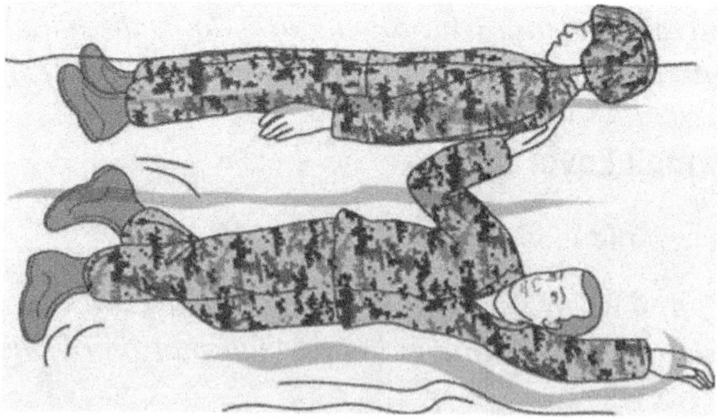

Once the victim is horizontal and on the water's surface, begin forward momentum by extending your towing arm to a fully locked-out position and executing a single armpit tow.

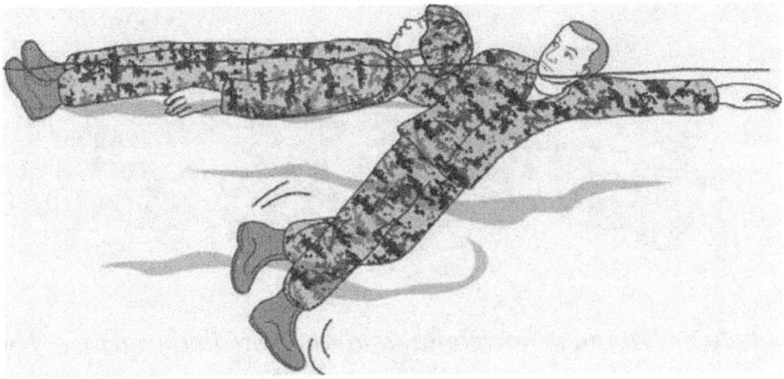

Double Armpit Level Off

Approach the victim.

Stop 2 to 6 yards from the victim to reassess the situation, reassure the victim, and maintain your margin of safety.

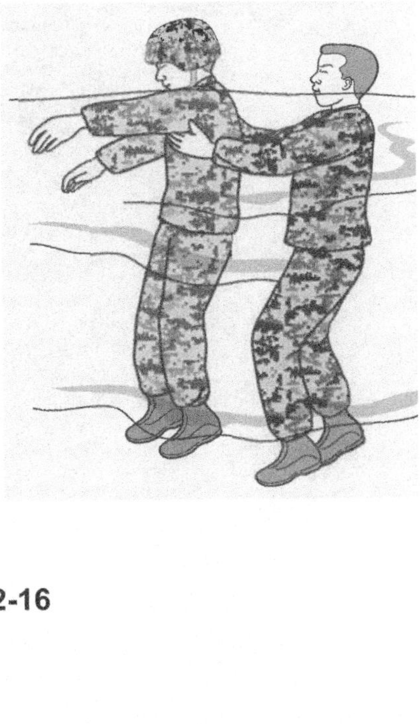

Approach the victim slowly and grab his armpits with both your hands. Place your elbows on his back.

Pull with your hands that are in the victim's armpits while pushing with your elbows that are in the victim's back and use an inverted breast stroke kick to place the victim horizontal on the water's surface (face up).

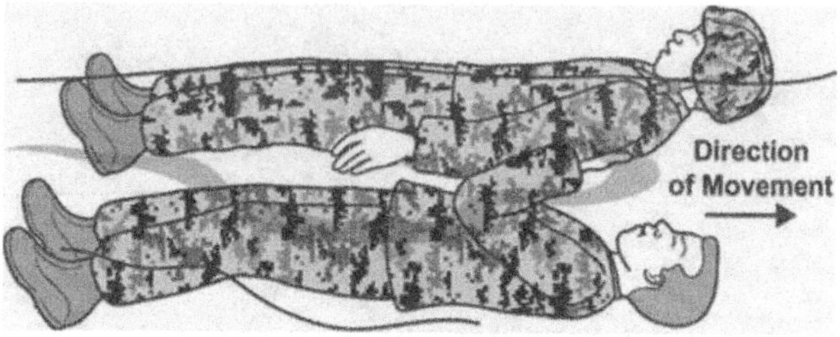

Once the victim is horizontal and on the water's surface, begin rearward momentum by slowly extending your arms into a fully locked-out position and executing a double armpit tow and inverted breast stroke kick.

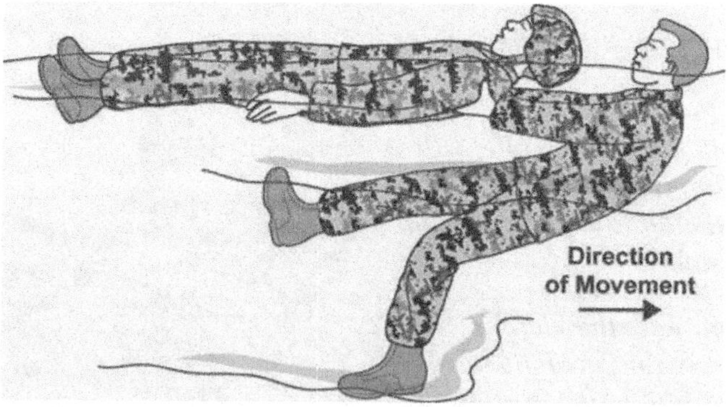

Rescue Techniques

Wrist Tow

Use the wrist tow method to rescue a victim who is floating face down. DO NOT use the wrist tow on a struggling victim. If time allows, remove your helmet and gear before attempting the rescue. Swim toward the victim using a modified breast stroke. Swim within 2 to 6 yards of the victim to maintain a margin of safety, this allows you to reassess the situation and reassure the victim. The following steps show proper front surface approach, wrist tow procedures:

Approach the victim from the front and grasp the underside of the victim's left wrist with your left hand or the right wrist with your right hand. Ensure that your thumb is on the underside of victim's wrist.

Lean back, pulling and kicking strongly to move the victim into a horizontal position.

Twist the victim's wrist to rotate the victim into a face-up position.

Swim toward safety using the lifesaving stroke.

> *NOTE:* The lifesaving stroke is a modified side stroke wherein the top arm is used to tow or carry a victim to safety and use a scissor kick or an inverted scissor kick is used for propulsion.

Keep a firm grip on the victim's wrist.

Keep your towing arm fully extended and along your side. This keeps the victim in the water column and prevents drag.

Ensure the victim's head does not go under water during the recovery.

Single Armpit Tow

You perform the single armpit tow after a victim has been properly leveled off. The single armpit tow uses the lifesaving stroke. To execute the single armpit tow—

Place one hand (top arm) in the victim's armpit. Your towing arm is straight and along your side.

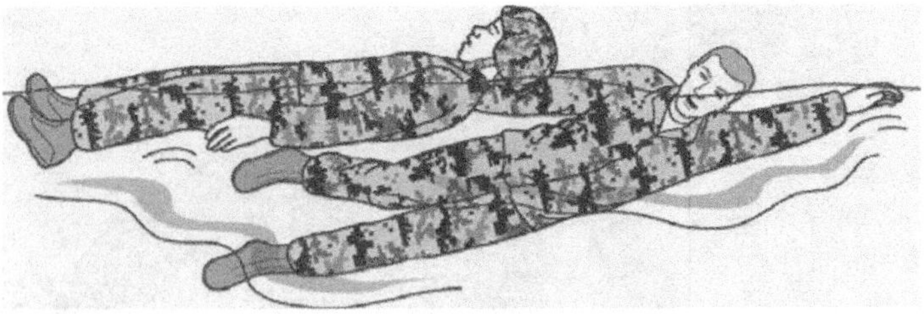

Use your lead arm (bottom arm) to execute short, vigorous pulls.

Execute a scissor or inverted scissor kick in a continuous and vigorous manner.

Use either free breathing or explosive breathing during the tow. During free breathing, keep your head above the water's surface and continuously reassure the victim. During

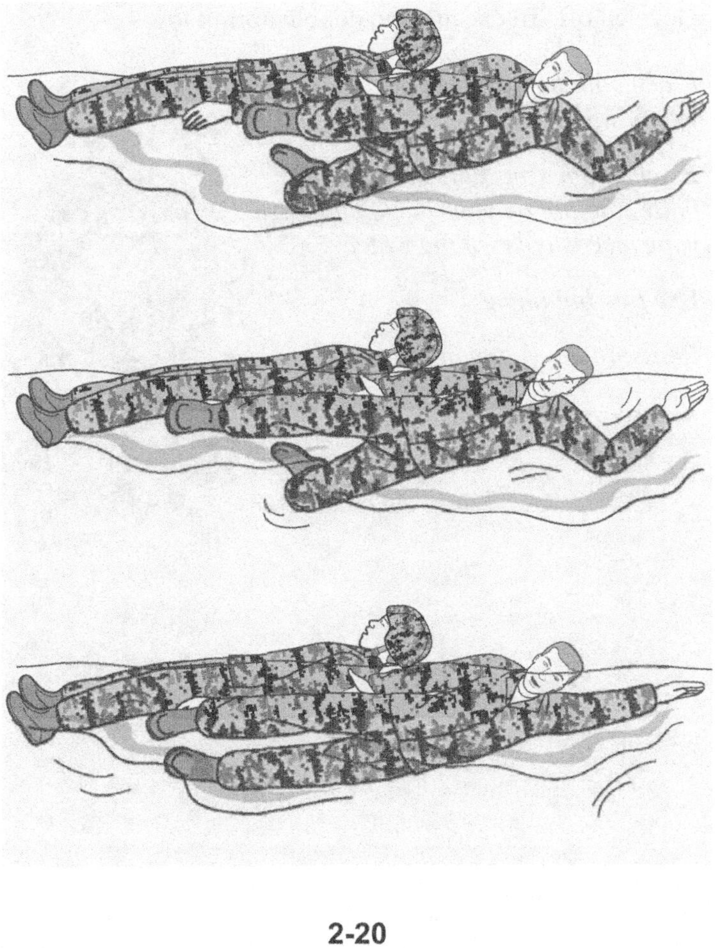

explosive breathing, put your head under the water to plane off your body's angle and reduce drag. Execute approximately two strokes, then lift your head up to breath and reassure the victim. Return your head back underwater for the next few strokes.

Double Armpit Tow

You perform the double armpit tow after a victim has been properly leveled off. To execute the double armpit tow—

Place both hands in the victim's armpits.

Extend both arms fully along your body. You are on your back, and your face is clear of the water.

Use free breathing.

Reassure the victim at all times.

Use the inverted breast stroke kick (it is the only kick that can be used due to your body position) to tow the victim to safety. (The inverted breast stroke kick is the same kick used in the elementary backstroke.) Your kick must be continuous and vigorous in order to keep the victim's face above the water.

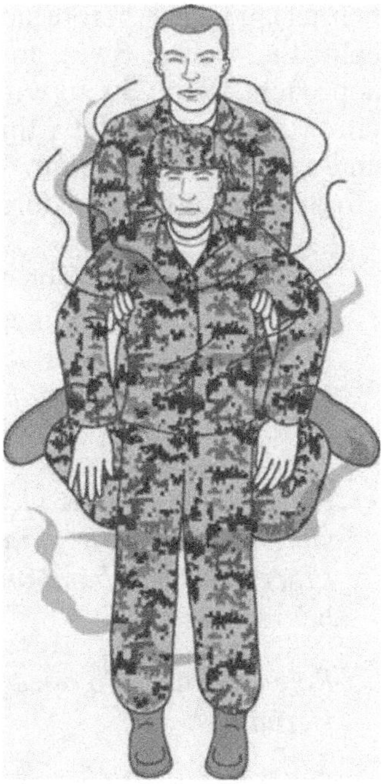

Collar Tow

You perform the collar tow after a victim has been properly leveled off. To perform a collar tow using the victim's blouse—

Maintain control of the victim by grasping his armpit with one hand and then with your free hand grasp either his combat gear or his blouse between his shoulder blades. If grasping the blouse, grasp the material with your palm up, then turn your hand over to tighten the material.

Release the victim's armpit once control is established and execute the lifesaving stroke to tow the him to safety.

Cross-Chest Carry

Use the cross-chest carry to carry a victim to safety if the victim is struggling or when moving through heavy surf. Remove your helmet and gear before attempting a rescue. Talk constantly to calm the victim. Swim toward the victim using a breast stroke approach stroke. Swim within 2 to 6 yards of the victim to maintain a margin of safety, this allows you to reassess the situation and reassure the victim. The following steps illustrate proper cross-chest carry procedures:

> **CAUTION:** The cross-chest carry causes fatigue even if you are in excellent physical condition.

Use a level-off technique to place the victim in a horizontal, face-up position.

Retain a grip on the victim with one hand. Reach over the victim with your free hand to encircle the victim's chest. Place your free hand on the victim's opposite rib cage, just below his armpit.

Release your grip once you have a secure hold on the victim's chest.

Swim toward safety using the lifesaving stroke while keeping a firm grip on the victim's chest and your hip on his back.

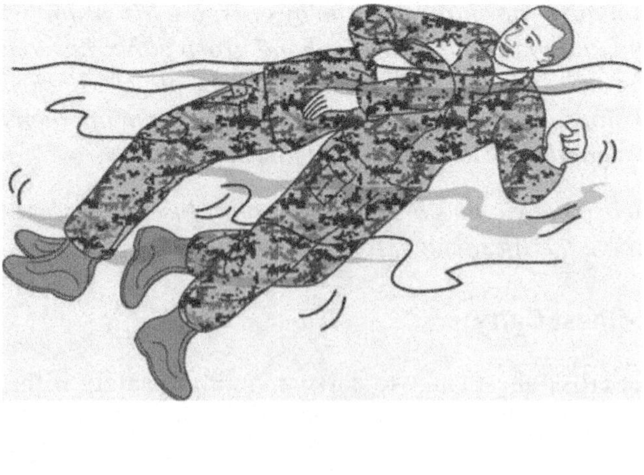

This procedure brings the victim's face and shoulders clear of the water, and typically the victim stops struggling. Sometimes, however, the victim will struggle during the swim to safety. If this happens, either tighten your grip on the victim or defend yourself with one of the techniques discussed on pages 2-25 through 2-30.

Tired Swimmer's Assist

You must maintain a 2 to 6 yards margin of safety from the victim at all times while you are getting into position to perform the tired swimmer's assist. Once behind the victim, you extend one arm, hand straight. Place your straight hand underneath the victim's opposite armpit. Maintain a 45 degree angle away from the victim, your arm is locked out. Assist the victim in propulsion until both of you reach safety.

Defense Against a Drowning Victim

DO NOT sacrifice your life in an attempt to save the victim. A struggling drowning victim poses great danger to anyone nearby. Driven by panic, the victim can grab you with great strength in an effort to climb out of the water. This can result in death for both you and the victim. The following techniques can be used to defend against a drowning victim, but the best defense against attack by the victim is to stay out of reach.

Block

The block prevents the victim from grabbing you if you have approached from the front. If the victim lunges toward you, react as follows:

Place one or both open hands against the victim's upper chest; being careful to avoid the victim's face, neck, and abdomen.

Lean backwards and submerge rapidly. Keep your blocking arm(s) extended.

Swim underwater and away from the victim, and return quickly to the surface.

Stop 2 to 6 yards from the victim to reassess the situation.

Determine an appropriate course of action.

Wrist-Grip Escape/Wrist-Grip Escape Alternative

The wrist-grip escape is used when a victim grabs your arm or wrist. Submerge the victim quickly by reaching across with your free hand, pushing down on the victim's shoulder to submerge him, and kicking to propel yourself upward. While keeping the victim submerged with your hand on his shoulder, give three hard jerking pulls with your trapped hand in an attempt to break free from his grasp. Once free, swim clear of the victim and reassess the victim's condition.

If the wrist-grip escape is unsuccessful in freeing your hand, use the wrist-grip escape alternative: use your free hand to grab your trapped fist, rotate thumbs up, apply bone-on-bone contact with the victim's arm, pry your hand out of his grasp, and quickly swim away from the victim.

Front Head-Hold Escape

The front head-hold escape allows you to escape when you are facing a victim who is gripping you around your head and neck. To execute the front-head hold escape—

Take a quick breath and tuck your chin into your shoulder to protect your throat.

Clap your hands above your head (three times) to submerge instantly. This drags the victim below the water, lifts his arms from around your neck, and, typically, he releases his grasp in order to get back to the surface. If he doesn't release his grasp, apply pressure to the victim's brachial pressure points (which are located inside of the upper arm, above the elbow).

Thrust the victim's arms up and away.

Keep your chin tucked to protect your throat, and swim underwater away from the victim and return quickly to the surface at the ready position.

Stop 2 to 6 yards from the victim to reassess the situation.

Determine an appropriate course of action.

Rear Head-Hold Escape

The rear head-hold escape allows you to escape when a victim is gripping your head and neck from the rear. To execute the rear head-hold escape—

Take a quick breath. and tuck your chin down, turn your head to either side, and raise your shoulders to protect your throat.

Take a strong stroke, clap your hands above your head (three times), and submerge instantly. This drags the victim below the water and, typically, he releases his grasp in order to get back to the surface. If he doesn't release his grasp, apply pressure to the victim's brachial pressure points (which are located inside of the upper arm, above the elbow).

Thrust the victim's arms up and away.

Twist your head and shoulders until free.

Swim underwater away from the victim and return quickly to the surface.

Stop 2 to 6 yards from the victim to assess the situation.

Determine an appropriate course of action.

Administering First Aid/Rescue Breathing

To administer first aid/rescue breathing, remove the unconscious victim from the water when possible and stop any bleeding.

First, open the airway and check for breathing. If the victim is not breathing, give two breaths of air. Check for a pulse. If there is a pulse but the victim is not breathing, continue rescue breathing. If there is no pulse, start cardiopulmonary resuscitation (CPR). For detailed information on administering CPR and rescue breathing, refer to American Red Cross CPR skill cards.

WARNING

If the victim has no pulse and is not breathing, administer CPR immediately. If the victim does have a pulse but is not breathing, give rescue breathing only. If the victim has a pulse and is breathing, DO NOT give CPR—CPR could prove fatal.

Next, protect any wounds from exposure, and treat for shock.

Chapter 3
Treatment of Casualties and Avoidance of Dangerous Marine Life

To survive in the water, you will face many challenges other than just being able to swim. For example, injuries and/or fatigue can lead to drowning and exposure to the elements can lead to hypothermia or heat injuries and may require medical attention. Finally, your presence in the water may be seen as a threat to or food source for a variety of marine life and you may be attacked. This chapter advises you on how to minimize these risks in order to stay alive.

Drowning

Drowning is suffocation by a liquid. A drowning victim inhales water into the lungs or the throat closes by reflex so that little or no water can enter the windpipe. In either case, a victim can no longer breathe.

Symptoms

One symptom of drowning is that the victim may call for help and has an expression of dread or panic. But typically a victim that is active and drowning may not call for help because he is trying to conserve his air and will not speak.

Another symptom of drowning is when the victim thrashes at the water's surface. If the victim stops or grows calmer, he has likely been overcome by fatigue, hypothermia, or a lack of air. At this stage, the victim usually has 20 to 60 seconds before going under the water's surface.

Treatment

If the victim is not breathing, begin rescue breathing. Place the victim on his back, tilt head back to open airway, pinch the nose, and give two full breaths. If the victim does not inhale during the first two breaths, reposition his head and attempt two more breaths. Check for a pulse. If a pulse is present, but the victim is still not breathing, continue rescue breathing. If a pulse is not present, begin CPR. See MCRP 3-02G, *First Aid*, for rescue breathing and CPR details.

WARNING

> If the victim has no pulse and is not breathing, administer CPR immediately. If the victim does have a pulse but is not breathing, give rescue breathing only. If the victim has a pulse and is breathing, DO NOT give CPR—CPR could prove fatal. Continue first aid until medical help arrives.

A victim who is not breathing and has no pulse may appear dead. DO NOT decide that death has occurred. Continue with the prescribed treatment. A corpsman or medical officer should decide whether the victim can or cannot be revived.

Hypothermia

Hypothermia is the abnormal lowering of the body's internal (or core) temperature to 95 degrees or below. It occurs when the body loses heat faster than it produces heat. A hypothermia victim loses the ability to move quickly, becomes mentally sluggish, slips into semiconsciousness, lapses into a coma, and dies when internal body temperatures drop too low.

The chilling effects of cold air, wind, or water can produce hypothermia. Water poses the greatest threat because it transfers heat 25 times faster than air. Depending on the water's temperature, a victim can succumb to hypothermia within a few minutes. The body's sudden contact with cold water can also set off a body reaction known as the mammalian diving reflex. This reflex can greatly increase survival time (especially for women and children) in or under cold water. The mammalian diving reflex shuts off blood circulation, except for the flow between the heart, lungs, and brain. The small amount of oxygen left in the blood and lungs is saved for the body's vital organs. This reflex has allowed people to survive being under cold water for an extended period of time. Therefore, a cold water drowning victim should be treated as if still alive even though the victim is not breathing, has no pulse, and may appear dead. DO NOT decide that death has occurred. Continue with the prescribed treatment. Victims of hypothermia can appear to be dead when they are not. A corpsman or medical officer should decide whether the victim can or cannot be revived.

WARNING

> If the victim has no pulse and is not breathing, administer CPR immediately. If the victim does have a pulse but is not breathing, give rescue breathing only. If the victim has a pulse and is breathing, DO NOT give CPR—CPR could prove fatal. Continue first aid until medical help arrives. Check for a pulse for at least 45 seconds.

Symptoms

Once the body's core temperature drops, the victim will show one or more of the following symptoms:

- Violent and uncontrollable shivering as the body tries to warm itself.
- Slow or slurred speech.
- Disorientation or poor coordination.
- Loss of skin color.
- Blue and pinched lips.
- A slowing or stopping of shivering that progresses into a rigid torso and limbs.

Survival Time

A hypothermia victim's survival depends on the water's temperature and the time spent in the water. A small body build cools faster than a large build. Children cool faster than adults.

To increase your chance of survival in the water, utilize the HELP position described on page 1-37. Extra clothing and inactivity (remaining motionless in the water) can also increase your survival time.

Treatment

A hypothermia victim must be warmed to prevent further heat loss; therefore, treatment should begin as soon as possible. Consciousness of the victim determines the treatment that should be pursued. See MCRP 3-02G for specific treatment information. The following treatment procedures are recommended:

- If the victim is conscious, give the victim warm fluids. Give candy or sweetened foods to a victim who is able to eat.
- If the victim is unconscious, place him on his back with his head tilted back to ensure an open airway.
- DO NOT massage the victim. Massage can break blood vessels and create swelling, internal pressure, and blocked blood circulation.
- DO NOT give alcohol to the victim. Alcohol lowers the victim's body temperature.
- Shock is a possibility, treat accordingly.
- Seek medical help immediately.

If you are able to remove the victim from the water, apply the following steps when possible:

- Get the victim into shelter.
- Remove the victim's wet clothing.
- Put the victim in dry clothing.
- Place the victim in a sleeping bag if one is available. It may be necessary to place another Marine in the sleeping bag with the victim.
- Place as much insulation as possible between the victim and the ground.
- Use hot water bottles, electric blankets, or blankets heated in an oven or by a campfire to warm the victim's neck, groin, and the sides of the chest.

CAUTION: DO NOT apply heat to extremities.

Heat-Related Injuries

Heat-related injuries include heat cramps, heat exhaustion, and/or heat stroke. They can occur when a Marine—

- Is exposed to extreme heat, such as from the sun or a combination of high air temperatures and water temperatures.
- Does not wear proper clothing or gets overexposed to the sun.
- Becomes dehydrated.

With heat cramps, muscles may cramp in the arms, legs, and/or stomach. The victim may sweat excessively. To treat, create improvised shade for the victim and have him drink water.

With heat exhaustion, the victim sweats heavily; presents pale, moist, cool skin; and complains of a headache, weakness, dizziness, and/or loss of appetite. A victim may also experience heat cramps, nausea, vomiting, an urge to defecate, chills, rapid breathing, confusion, and a tingling sensation in the hands or feet. To treat, pour water over the victim and fan him to speed up the coolant effect of evaporation, have the victim drink freshwater, and attempt to provide shade.

With heat stroke, the victim stops sweating, which results in red, flushed, hot, dry skin. The victim may first experience headaches, dizziness, nausea, fast pulse and respiration, and/or seizures and mental confusion. The victim can become unconscious and die if not treated quickly. To treat, create improvised shade for the victim. Have him drink water and elevate both legs if possible.

Burns

Apply a field dressing (or the cleanest material available) to the burn. Give sips of water to a casualty who is conscious and not nauseated. When treating burn casualties—

- DO NOT remove clothing stuck to the burns.
- DO NOT break any blisters.
- DO NOT apply grease or ointments to the burns.

For electrical burns, check for both an entry and exit burn from the passage of electricity through the body. An exit burn may appear on any area of the body, not necessarily opposite the entry burn.

For burns caused by wet or dry chemicals, flush the burns with large amounts of water and cover with a dry dressing.

For burns caused by white phosphorus, flush the area with water, then cover with a wet material, dressing, or mud to exclude air and keep the white phosphorus particles from burning.

For laser burns, apply a field dressing.

Common Medical Problems Associated with Sea Survival

Seasickness

Seasickness is the nausea and vomiting caused by the bobbing motion created by the wave action of a flotation device. Seasickness can result in—

- Dehydration and exhaustion.
- A loss of the will to survive.
- Others becoming seasick.
- Unclean conditions.

To treat seasickness—

- Wash both the Marine and the flotation device to remove the sight and odor of vomit.
- Keep the Marine from eating food until the nausea is gone.
- Have the Marine lie down and rest.
- Give the Marine seasickness pills if available. If the Marine is unable to take them orally, the pills should be inserted rectally for absorption by the body.

Saltwater Sores

Saltwater sores occur when skin that has abrasions or is cut is exposed to saltwater. The sores may form scabs and pus. Do not open or drain. Flush the sores with freshwater, if available, and allow to dry. Apply antiseptic, if available.

Blindness/Headache

Irritants or the effects of the sun's rays reflecting off the water can cause temporary blindness or headaches. If flames, smoke, or other irritants get in your eyes, flush the eyes immediately with saltwater, then with freshwater, if available. Apply an ointment, if available. Bandage both eyes for 18 to 24 hours or longer if the damage is severe. If glare from the sky and water causes your eyes to become bloodshot and inflamed, bandage the eyes lightly. Try to prevent this problem by wearing sunglasses or goggles with a sunglass insert.

Constipation

This condition is a common problem associated with dehydration. For constipation, do not take a laxative if it is available; this causes further dehydration. Drink freshwater, if available.

Sunburn and Dehydration

The sun's rays reflect at all angles off the waves of the water; therefore, sunburn and dehydration are serious problems in sea survival. Try to prevent sunburn by—

- Erecting an improvised canopy, with available floating materials, to provide shade.
- Wearing your soft cover or using a cloth, such as a handkerchief, to cover your head.
- Covering your skin with sunscreen or lip balm from your first aid kit. Your lips, nostrils, eyelids, the backs of your ears, and the skin under your chin sunburns easily. If enough sunscreen cream is available, all exposed skin should be covered.

Dehydration is caused by the loss of the body's vital fluids. Dehydration in saltwater may result from a combination of factors such as a lack of water, the effects of saltwater on skin tissue, sunburn or vomiting from seasickness, and other causes. Sleep and rest and reduced water and food intake are the best ways of enduring periods of exposure. The following measures will delay the effects of dehydration:

- DO NOT drink saltwater.
- DO NOT drink urine.
- DO NOT drink alcohol.
- DO NOT smoke.
- DO NOT EAT unless water is available.

Dangerous Marine Life

You may see many types of marine life around you, and some are more dangerous than others. Generally, sharks are your greatest danger, followed by barracudas. However, most marine life will not deliberately attack a human. The most common injuries from marine life are wounds from bites, stings, or punctures. With the exception of sharks and barracudas, most injuries are a result of either trying to catch game or from contact abrasion with marine life. To treat an injury that results from a dangerous marine animal bite—

- Pack the wound with gauze and then apply a pressure bandage, if available.
- Treat for shock.
- Obtain medical attention as soon as possible.

For a more detailed information on the treatment of casualties and dangerous marine life, see MCRP 3-02F and MCRP 3-02G.

Sharks and Barracudas

Only about 20 percent of all shark species are known to attack people. Sharks have an acute sense of smell, and the smell of blood in the water will draw them to their prey. They are also very sensitive to any abnormal vibrations in the water; therefore, the sound caused by a struggling swimmer or underwater explosions will attract them.

Barracudas are bold and inquisitive fish. They have been known to attack men who are wearing shiny objects. Barracudas may charge at lights or shiny objects at night.

A group of swimmers can maintain a 360 degree watch while in the water. Therefore, to protect yourself from sharks and barracudas, stay with other swimmers. The group members can either frighten away or fight off sharks or barracudas better than one person can. Keep all your clothing on, including footwear. Historically, sharks attack unclothed individuals in groups first, mainly in the feet. Clothing also protects you against abrasions from a shark's tough skin should the shark brush up against you. Avoid urinating heavily, let urine dissipate between discharges. If you must defecate, do so in small amounts and toss it as far away from the group as possible. Do the same if you must vomit.

If attacked, the use of firearms by swim sentries should be used with extreme caution because of the risk of injury to other swimmers. If unarmed or unable to make an improvised weapon, kick and strike the shark. Avoid using your bare hands to strike the shark, injury can result to your hands due to a shark's tough skin. Target areas on a shark are the gills, eyes, and underbelly. Blows to the snout are also not recommended because a shark will tilt its head up and thrust its jaws forward when biting.

Sea Snakes

Sea snakes are venomous and sometimes found in mid-ocean. They are unlikely to bite unless provoked. Your best protective measure is to avoid them.

Poisonous Fish

Many reef fish are toxic, and can kill you if eaten. Generally, poisons are present in all parts of the fish, but especially concentrated in the liver, intestines, and eggs.

Turtles and Moray Eels

Turtles and moray eels normally inflict minor bite wounds. Treat this type of injury by cleaning the wound.

Corals

Coral, dead or alive, can inflict painful cuts. Clean all coral cuts thoroughly. DO NOT use iodine to disinfect any coral cuts. Some coral polyps feed on iodine and may grow inside the flesh.

Jellyfish, Portuguese Man-of-War, Anemones, and Others

This group of marine animals inflicts injury by stinging their victims with their tentacles. Contact with their tentacles produces burning pain, a rash, and small hemorrhages on the skin. Shock, muscular cramping, nausea, vomiting, and respiratory distress may also occur. Gently remove the clinging tentacles with a towel and wash or treat the area. Use diluted ammonia or alcohol and talcum powder to treat the injury if available.

Spiny Fish, Urchins, Stingrays, and Cone Shells

These animals inject their venom by puncturing the skin with their spines. General signs and symptoms include swelling, nausea, vomiting, generalized cramps, diarrhea, muscular paralysis, and shock. Deaths are rare. Treatment consists of soaking wounds in hot water, if available, to deactivate heat-sensitive toxins.

Chapter 4
Negotiating Water Obstacles

Marines face water obstacles in saltwater, freshwater, and brackish water (where freshwater and saltwater meet). These water environments differ considerably and pose distinct problems for Marine tactical units and swimmers. Saltwater obstacles include tides, surfs, and currents. Brackish water obstacles include back bays. Freshwater obstacles include rivers and canals.

Tides

Tides are periodic changes in the surface levels of oceans, bays, gulfs, inlets, and rivers. The Moon's and Sun's gravitational pulls cause tides. Depending on the situation, tides can either help or hinder Marines in their endeavors to conduct amphibious or riverine operations. Tides can provide sufficient depths to allow for the passage of landing craft or boats over reefs, trees, and other underwater obstructions. On the other hand, tides can render a river fording site unusable to tactical vehicles. Direction, level of change, and amount of change determine tidal nomenclature.

Tides that show change in direction are flood tides and ebb tides. Rising tides are known as flood tides. Falling tides are known as ebb tides.

Tides that show extreme levels of change are high tides and low tides. High tide is the period when water is at its greatest depth. Low tide is the period when water is at its most shallow depth.

Tides that show the least and most amount of change are neap tides and spring tides, respectively. Neap tides have the least amount of change in water levels between high and low tide. Neap tides occur at the half moon, when the Sun and the Moon are aligned at a 90-degree angle with the Earth. In this position, the Sun's and Moon's gravitational pulls offset each other. Spring tides have the highest floods and lowest ebbs. Spring tides occur at or shortly after the new moon or full moon when the Sun, Moon, and Earth are approximately in line. In this position, the Sun's and Moon's gravitational pulls are combined.

Surf

Waves break upon entering shallow water and create surf. The offshore area where waves break is the surf zone, which can present many hazards. Breaking waves often trap air bubbles and create a foamy appearance. Bubbles lower the water's density and decrease buoyancy. Move through foamy surf as quickly as possible. The type of wave determines your survival technique.

Wave action moves you toward shore. Lie on your back or side with your head pointing in the direction of the beach and your feet pointing into the waves.

As one wave approaches the beach, another drains away; relax and do not swim against the draining water. When a new wave is within about 3 yards, start swimming toward shore. Continue to swim until the wave lifts you and moves you toward the beach. Once the wave loses forward momentum, relax and repeat the cycle. If nearing rocks, turn your body and approach feet first to reduce the chance of striking your head and arms.

Plunging Waves

A plunging wave is a breaker whose crest curls forward and falls ahead of its base. Because of its power and underwater turbulence, a plunging wave poses the greatest surf threat. If caught in a plunging wave, you can be pulled underwater and pitched about violently. This can cause you to panic, which can increase your chance of drowning. Perform the following steps to escape a plunging wave:

Tuck into a ball with your head against your knees and your forearms locked around your legs, just below the knees.

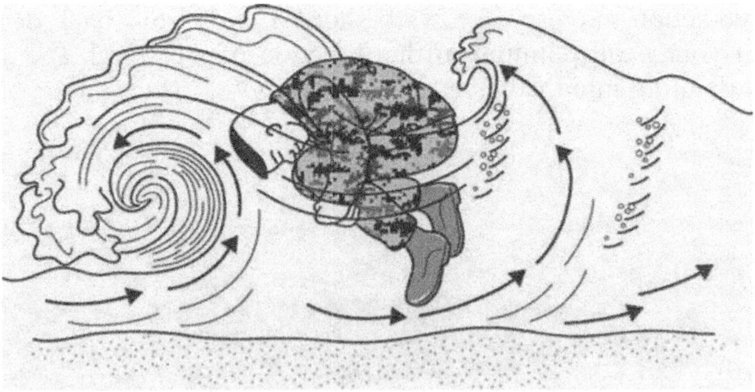

Relax in this position until the turbulence subsides and you float to the surface. This can take 30 seconds or more.

Swim toward shore.

> *NOTE:* If threatened by another plunging wave, dive underwater into the wave.

Spilling Waves

A spilling wave does not break. Instead, its crest slides forward without curling. A spilling wave creates less turbulence and poses less of a threat than a plunging wave. If caught in a spilling wave, relax and let the wave carry you to shore.

Surging Waves

A surging wave occurs on a beach with a steep underwater gradient. It never really breaks, but the crest rises while the base slides up the beach with great force and speed. Once the wave reaches its highest point on the beach, it rushes back as quickly as it surged forward. If you are standing on the bottom when a surging wave advances or retreats, the wave can knock you off your feet and pull you into the surf zone. Do not try to stand or walk on the bottom. Swim toward the beach as soon as possible.

Currents

Offshore Currents

An offshore current occurs outside the surf zone. Typically, it occurs at bay entrances, in island channels, and between islands and the mainland. An offshore current flows parallel to or away from shore. If the offshore current is created by tides, its current strength and direction vary at different times of the day.

If caught in an offshore current, you may be carried in a direction you do not want to go. DO NOT try to swim directly to safety. If the current is moving directly away from the shore, relax and wait until the current dies out or turns toward land. Once the current subsides, use any survival stroke to swim toward shore. If the current is moving parallel to shore, use any stroke to move at an angle across the current and toward shore.

Rip Currents

A rip current occurs when waves pile water against the shore faster than the water can drain. The water flows rapidly along the beach until it is deflected seaward by a bottom obstruction. Then the rip current flows through the surf zone and into open water at a speed of up to 2 knots. This action can cut deep trenches in the sand. A rip current dies out once it reaches open water (usually within a few hundred yards of the shore).

A rip current can pull you out to the open sea. If caught in a rip current, DO NOT try to swim against the current. A rip current moves faster than most people swim, and it is impossible to swim to shore once caught in it. Relax and stay afloat until the current runs out. Once the current subsides, use a survival stroke to move parallel to the shore until you are out of the current, then begin swimming toward shore.

Littoral Currents

A littoral current occurs when a wave breaks against a beach at an angle. This current flows parallel to the shoreline and does not pose a great threat. If caught in a littoral current, use the combat travel stroke to swim across it at an oblique angle.

Back Bays

Once on the beach, you often face one or more rows of low hills called dunes. Behind the dunes, you may encounter a low-lying stretch of ground thickly covered with scrub trees and bushes. This area gives way to wetlands known as back bays. Back bays consist of muddy islands that are almost submerged during flood tides and separated by channels of brackish water of varying depths. Channel bottoms usually contain soft mud. Back bays pose major obstacles to vehicular traffic.

Infantry can cross back bays, but only with great effort. If crossing back bays by foot, consult detailed navigation charts and use the following guidelines to plan your route:

- Avoid water less than waist deep; walking in shallow water or soft mud is extremely tiring.
- Avoid back bay islands; these low-lying islands are usually too muddy to support foot traffic.
- Seek out deep water; floating with a pack is less tiring than walking through shallow water or soft mud.
- Seek out sand, shell, gravel, or stone bottoms; these firmer bottoms generally ease travel and help conserve energy.

Rivers and Canals

A river is a large, natural stream of water that empties into a larger body of water. The slope of the riverbed and the volume of water in the river determine its current.

Canals resemble small rivers or streams in their width and depth, but usually lack any significant current. Climbing out of these waterways can be difficult if the canal is flanked by steep banks.

Chapter 5
Fording Waterways

WARNING

Fords are dangerous. Cross as quickly as possible.

A ford is any site in a river, stream, or canal where the water is shallow enough for troops or vehicles to cross without using flotation devices. Canal bottoms are usually too soft to support fording vehicles, and wading infantry frequently stumble.

The tactical situation dictates the location of the fording site. Seek fords that are protected from enemy observation and that allow for adequate supporting fires. A night fording takes at least one and a half times as long as a daylight fording.

> *NOTE:* In this chapter, the term "river" refers to rivers, streams, and canals.

Selection of a Ford Site

The following table identifies desirable fording characteristics:

Characteristics	Comments
Concealment	The ford hides personnel and vehicle movement from enemy observation.
Accessibility	The ford should have low banks with gentle gradients. This allows a free flow of traffic at both the entrance and the exit.

Characteristics (Continued)	Comments (Continued)
Slow Current	The ford's current should not exceed 1.5 meters per second if possible.
Firm Footing	The ford's bottom, entry, and exit composition should be firm enough to support traffic. Do not drive a vehicle over any bottom composition that a 2-inch diameter stick can be pressed into more than 1 or 2 inches.
Gently Sloped Channels	The ford's entry and exit points should be gently sloped. If possible, locate a portion of the stream where the channel is not actively shifting.
Depth	The fording depth is less than or equal to the least capable vehicle.

Determine the Slope

Units move into and out of water faster and more quietly if entry and exit points are not steep or muddy. Slope is the amount of change in ground horizontal distance (run) and in vertical elevation (rise) from one point to another. Slope is usually expressed as a percentage. You can use a clinometer, map, or line of sight and pace to measure the percentage of a slope.

Clinometer

A clinometer measures percentage of a slope. It is a component of the M9 armored combat excavator, and organic to many Marine engineer units. However, engineers are also taught (and often use) field-expedient methods to determine slopes and gradients.

Map

A map measures the horizontal distance along a desired path. Determine the difference in elevation between the path's starting and ending points. From a map's scale, you can determine the distance between two points. From a map's elevation lines, you can determine the difference in height between the same two points. Both figures must be in the same unit of measure (e.g., feet, meter). Divide the elevation (rise) by the distance (run) and multiply by 100.

Rise = 165 - 120 = 45 meters

Run = 200 meters

% Slope = $\frac{45}{200}$ **x 100 - 22.5%**

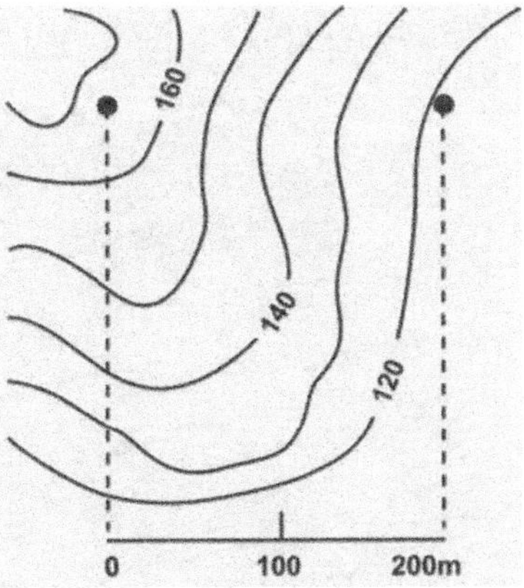

Line of Sight and Pace

To determine line of site and pace, stand at the bottom of the slope, keep your eyes level, pick a spot on the slope, then pace the distance. The number of paces multiplied by a standard measure of 0.75 meter determines the run. The eye-level height (usually 1.5 to 1.7 meters) determines the rise. Repeat this procedure until you have covered the entire distance you want to measure for each spot (vertical and horizontal). Add the vertical distances to provide total rise and the horizontal distances to provide total run.

Determine the Current Speed

Current speed increases as channels narrow. It may be necessary to locate a wider ford location to obtain a slower stream current. The following steps are used to calculate the speed of the current:

Determine points A and B along a channel. Then measure the distance between those two points.

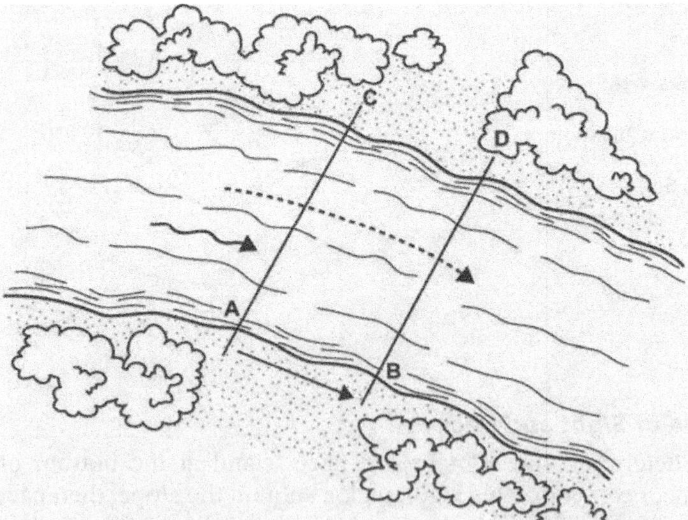

$$\frac{\text{Distance (meters) Between Points A and B}}{\text{Time (seconds) Between Points C and D}} = \text{Current Speed (meters per second)}$$

Sight directly across the water from points A and B to locate points C and D.

Throw a floating object (e.g., a stick) upstream from points A and C. Observe the object as it floats toward points B and D.

Subtract to find the time it takes for the object to float from start to finish.

Do not attempt to swim across currents that are moving faster than 1.5 meters per second. Equivalents of this speed include—

- Quick-time march rate of 120 counts per minute with one, 30-inch step at each count.
- 5 feet per second.
- 3.5 miles per hour.
- 5.5 kilometers per hour.

Measure River Width

A river's width can be estimated from the width of its symbol on a scaled topographic map. If this is not possible, use the following compass techniques:

Stand at the water line (A).

Shoot an azimuth to a point on the opposite bank (B).

Move upstream or downstream until you are at a point (C) where you can shoot an azimuth 45 degrees larger or smaller than the original azimuth.

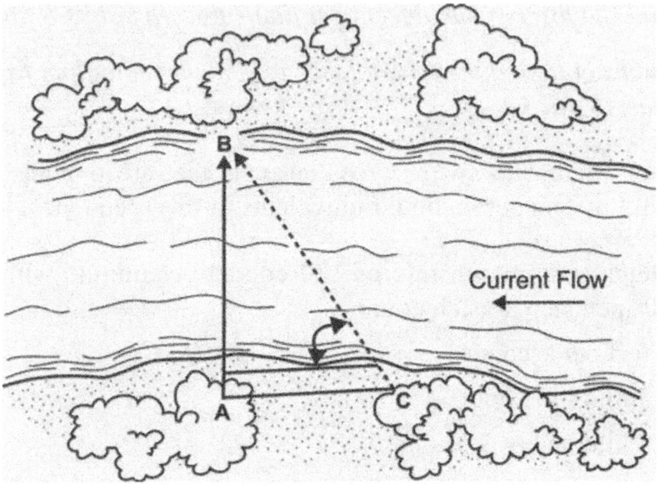

Measure the distance between points A and C. The distance calculated equals the river's width.

Calculate Downstream Drift

A river's current causes personnel and equipment to drift downstream. If personnel and equipment are aimed straight across the river, they will sideslip downstream as they move across the current to the other shore. Therefore, personnel and equipment crossings must compensate for the effects of a river's current and entries are usually made upstream of the desired exit point. Use the formula below to calculate downstream drift.

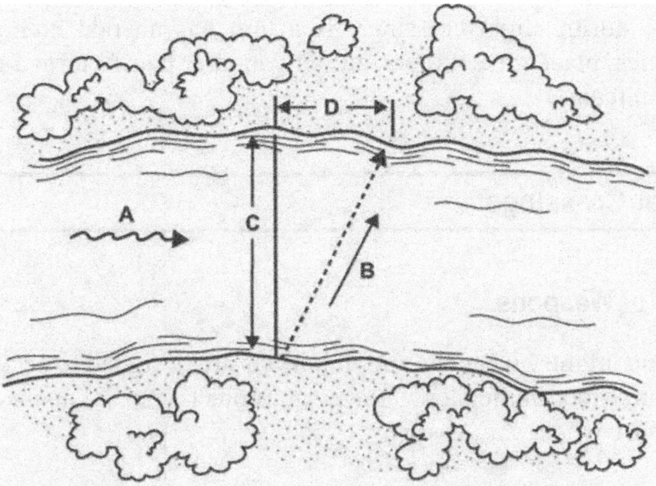

$$\frac{\text{Current Speed (A)}}{\text{Crossing Speed (B)}} \times \text{River Width (C)} = \text{Downstream Drift (D)}$$

> *Note:* The crossing speed for a swimmer across a river may vary but is generally limited to 1 meter per second. All measurements must be in the same unit of measure (e.g., meters, feet).

The Buddy System

Whenever a Marine unit must enter into or operate on the water, a "buddy system" is employed in which every Marine is paired with a swimming partner. The buddy system matches an experienced swimmer with a weak swimmer. The experienced swimmer assists and encourages the weaker swimmer and bolsters confidence during night crossings. If a unit has an odd number of Marines, place the extra person with another pair to form a three-person team.

Water Crossings

Care of Weapons

Marine infantry weapons and munitions are designed to be able to operate after immersion. However, protect your weapons from moisture whenever possible.

A gas-operated weapon can malfunction if water travels down the barrel and enters the gas tube. To protect the gas tube—

Close the weapon's bolt before entering the water.

Seal the muzzle with a condom, balloon, plastic spoon wrapper, or other form of waterproof material.

Tie or melt the protective cover to create a watertight seal.

When the muzzle's protective cover is no longer needed, remove it. Open the bolt and inspect the barrel. If the tactical situation permits, swab excess moisture from the barrel. Test fire automatic weapons, if possible. Field strip and clean weapons as soon as possible. If time does not allow for a complete inspection, rinse inaccessible areas with small amounts of diesel fuel, then dry.

ISOMAT Raft

Construction of an ISOMAT raft is time-consuming. This type of raft should not be employed as part of an attack, but used for logistical purposes (e.g., evacuating stretcher cases, transporting supplies). Use the following steps to build an ISOMAT raft:

Wrap ISOMAT sleeping pads around sturdy sticks.

Use parachute cord and square knots to tie the pads securely in place and to lash stick ends together in a rectangle.

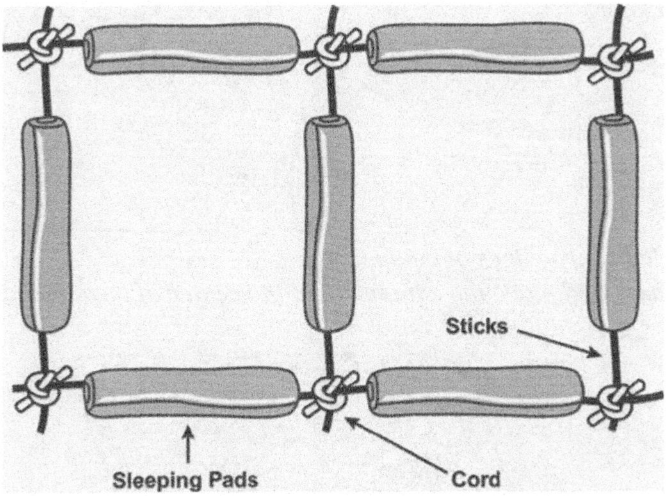

Note: The ISOMAT raft pictured can support several hundred pounds. However, the cargo will get wet if not properly waterproofed.

Poncho Raft

A poncho raft can support two Marines and their equipment and is well suited for long crossings. Use the following steps to build a poncho raft:

Inspect two ponchos and ensure they are serviceable.

Lay one poncho flat on the ground, with the hood-side up.

Cinch the hood tightly to form a gooseneck or tie in a knot.

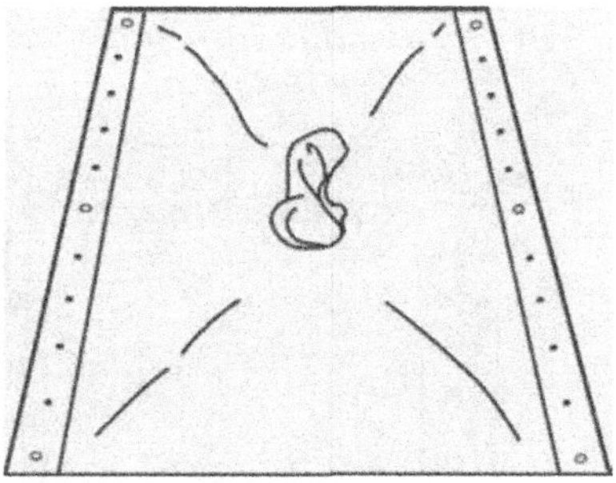

Pad sharp edges of equipment and place the equipment in the center of the poncho.

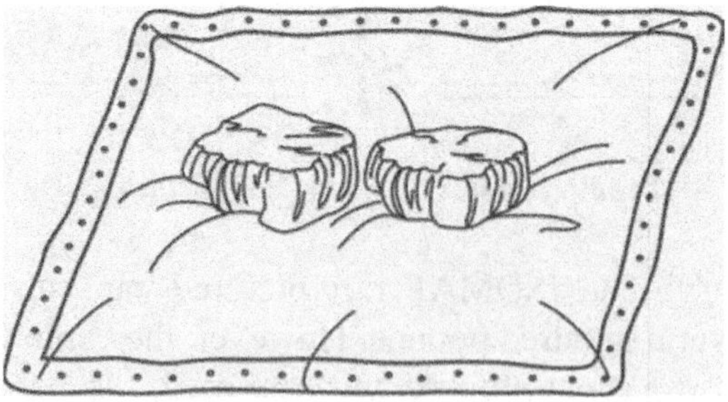

Place the second poncho over the equipment, rubber side up, and hood facing down.

Snap the edges of the two ponchos together.

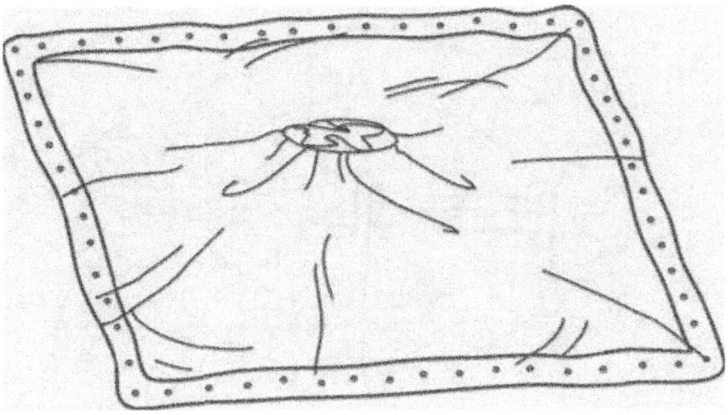

Roll the edges toward the equipment.

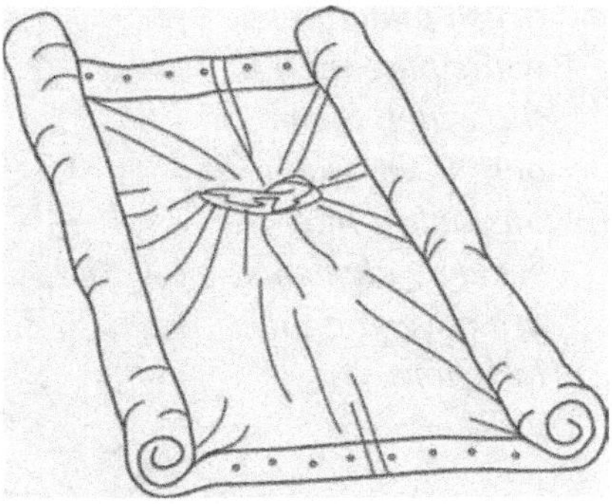

Roll the edges into pigtails and tie them off.

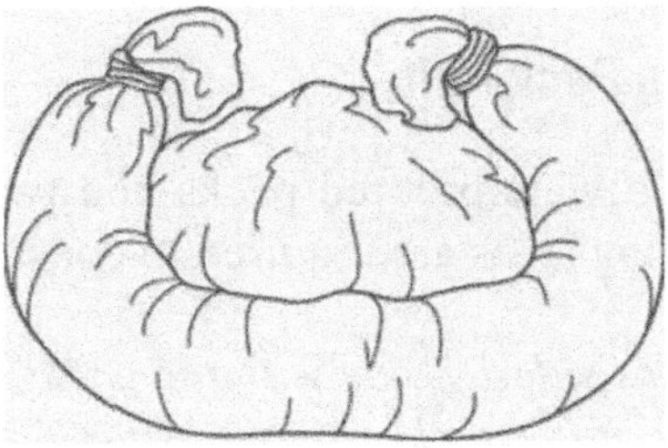

Pull the pigtails together over the top and lash them securely.

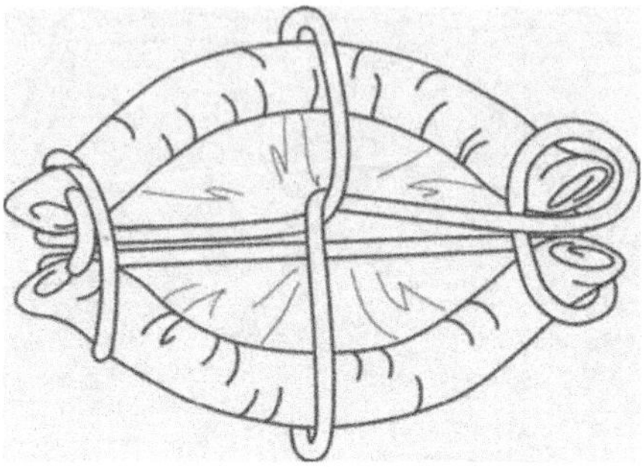

Protect the raft from brush punctures while placing it in the water. Swim across the water obstacle while security elements are covering the far shore.

Construction of a Pack Raft

You will need two waterproofed packs and two M16A2 service rifles. The following steps are required to construct a pack raft:

Place two packs side-by-side with the pack frames on the deck. The tops of the packs are opposite of each other.

Loosen the main compartment straps on both packs.

Insert one rifle on each end between the straps and the packs. The muzzles are opposite of each other. The rifles serve as one means to secure the packs together. Place the front sight post under the top flap.

Tighten the straps so that the rifles and packs are secure.

Take the excess strap on the inner side of each pack and secure it to the opposite pack to better secure the two packs together.

Take the excess straps on the outer sides of the packs and use those straps as safety lashing for the rifles.

Tuck the excess straps and check to make sure the rifles and packs are secure.

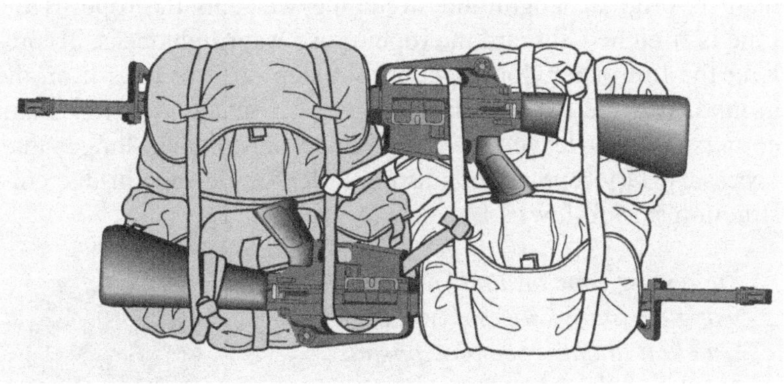

Single-Rope Bridge

A single-rope bridge offers a temporary and quick way to cross small rivers. It also provides extra security while crossing swift waters. At night, it prevents straggling, and guides units precisely from one side of the river to the other side. If crossing a river at night, plan for at least one single-rope bridge.

If your unit is crossing a river with swift currents or water depths above 4 feet, the unit is carrying sufficient rope to span the crossing site, and the tactical situation permits, secure the rope on near and far banks to provide a hand-hold for crossing Marines. This reduces the time required for the entire unit to cross and provides a degree of comfort/confidence for poor swimmers. Use a squad-sized bridge team to construct a single-rope bridge. Station several strong swimmers at the water's edge to help anyone who has trouble crossing.

Nylon rope is normally coiled in 120 foot lengths. It is 0.6 inches in diameter and has a breaking strength of about 3,840 pounds. Over time, a nylon rope can stretch to as much as one-third more than its original length and stretching weakens the rope. If the rope is stretched, discard the rope or use it for light tasks. To prolong the life of a nylon rope, do not step on it or drag it on the ground. Pad the rope in places where it contacts rocks or sharp corners. Do not leave the rope knotted or stretched longer than necessary. Dry rope as soon as possible. Single-rope bridge construction is as follows:

Tie a sling rope around your waist using a square knot and two, separate half hitches. See the appendix for detailed information on knots.

Attach a locking steel carabiner to the sling rope.

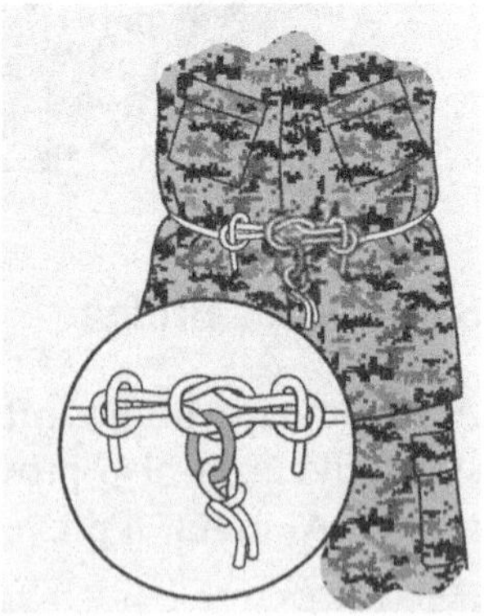

Tie a bowline knot in the running end of the bridge rope and attach it to the carabiner.

Temporarily secure the other end of the rope to a tree on the near shore.

Enter and cross the water.

NOTE: Carry only your weapon and ammunition.

Exit the water on the opposite shore.

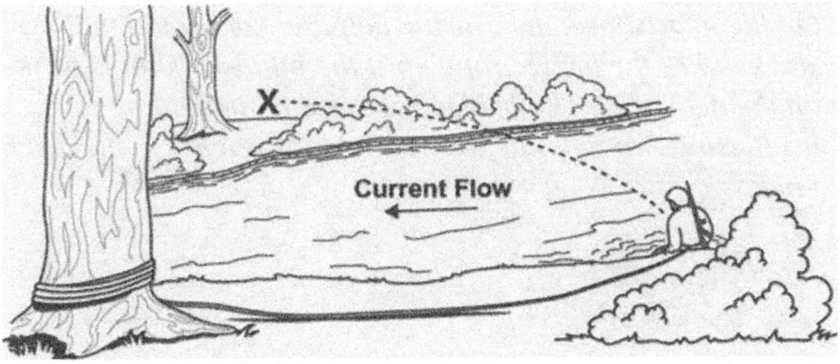

Prepare your weapon for use. Unhook the bridge rope from the carabiner at your waist, and tie the bridge rope to a sturdy tree using a round turn and two half hitches.

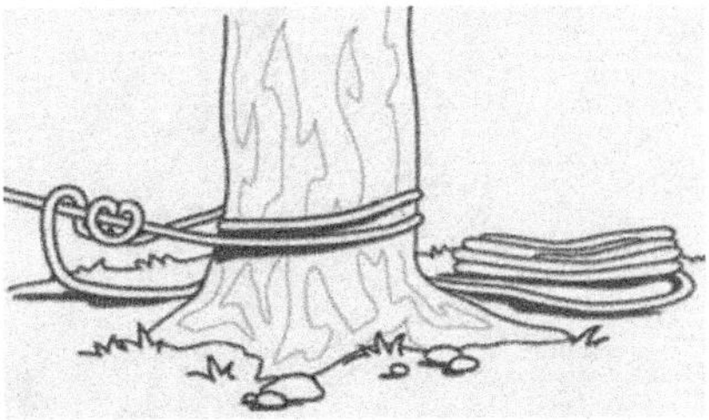

Conduct a box reconnaissance of the opposite shore.

On the near shore, have another Marine prepare to tighten the rope. That Marine should place a transport tightening system in the bridge rope by tying a double butterfly knot and placing two carabiners in the butterfly. See the appendix for detailed information on knots.

The Marine should pass the running end of the bridge rope around the downstream side of the near shore anchor point and through the two carabiners.

Pull the butterfly knot approximately one-third of the distance across the river.

Secure the bridge rope to an anchor point using a round turn and two half-hitches.

On the near shore, the Marine helping you should pull the slack out of the bridge rope until the butterfly knot is back on the near side. The bridge rope is then tied off against itself using two half hitches with a quick release in the last half hitch.

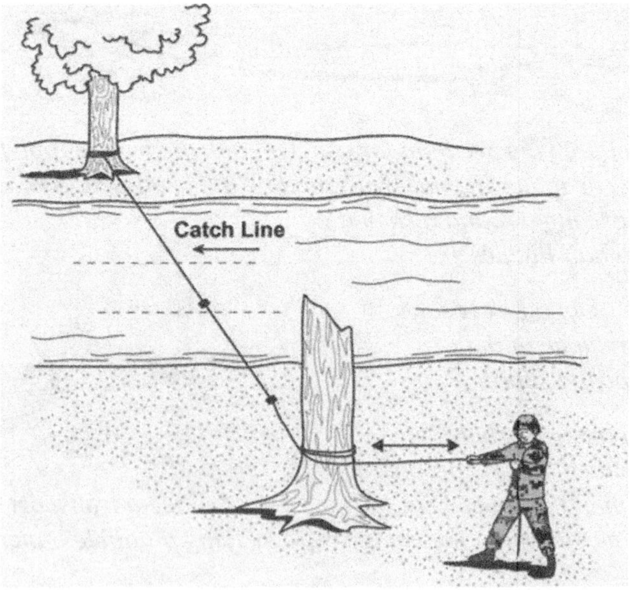

Note: The single-rope bridge must be as tight as possible so it will not sag when used.

If you lose your footing and fall into the water, swim with the current to the closest shore. Swimming against the current is dangerous and quickly causes fatigue.

High and Dry Crossings

If the single-rope bridge is high enough, suspend yourself below the single-rope bridge and above the water. Use the following steps to suspend yourself from a single-rope bridge and then pull yourself across the water:

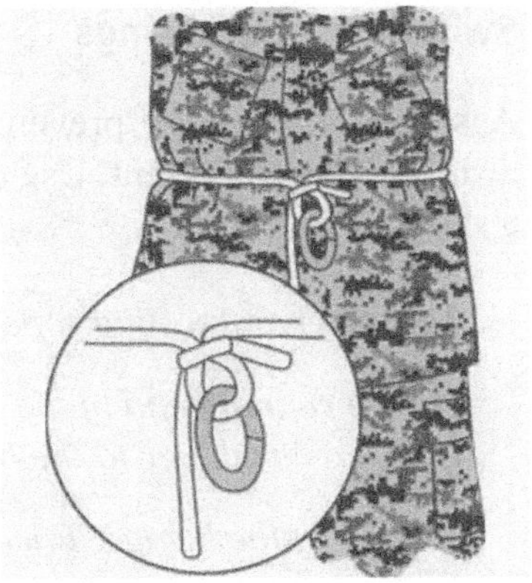

Tie a sling rope around your waist using a bowline. Ensure that the knot is tight.

Attach a carabiner through the bowline's loop. The carabiner's gate faces up.

Secure your helmet chin strap.

Face the single-rope bridge with your left shoulder toward the far shore.

Grasp the bridge rope in both hands.

Swing your body beneath the single-rope bridge with your head toward the far shore. Cross your ankles above the bridge rope.

Arch back until the carabiner contacts the bridge rope. Connect the carabiner to the bridge rope. Allow the carabiner to bear your body's weight.

Pull yourself across the single-rope bridge, hand over hand, to the far shore.

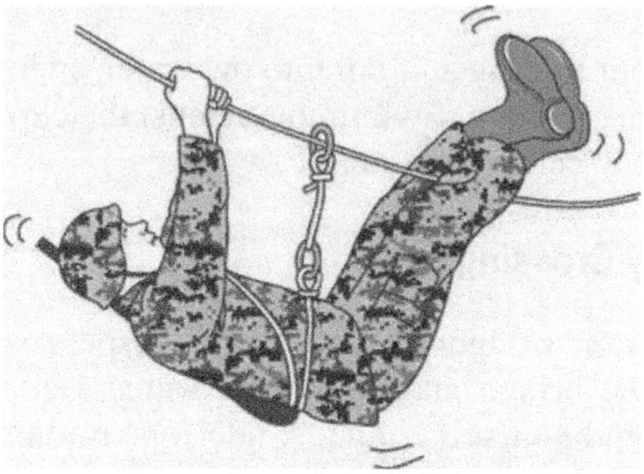

Swift Current Crossings

A single-rope bridge prevents being knocked down and swept away by a swift current. Use the following steps to move through a swift current:

Tie one end of a sling rope around your waist using a bowline.

Tie the running end of the sling rope in another bowline, and attach a carabiner to the bowline's loop.

Step up to the bridge. Face upstream.

Hook the carabiner to the single-rope bridge.

Walk sideways into the river while grasping the bridge rope in both hands.

Use the single-rope bridge for balance and remain standing, if possible.

Continue to move sideways through the river to the far shore.

Slow Current Crossings

If you face little or no current, it is not necessary to hook up to a bridge rope with a carabiner. Lie on your back in the water beneath the single-rope bridge. Support your body weight with your waterproof pack. Use the bridge rope and pull yourself hand over hand across the river.

Removal

If you are the last Marine waiting to cross, pull on the standing end of the rope to release the knot, then tie the rope around your waist using a bowline. The Marines on the far shore will pull you through the water.

Appendix
Knot Tying

Square Knot

The square knot is used to secure two ropes of equal diameter together so they form one continuous rope that will not slip.

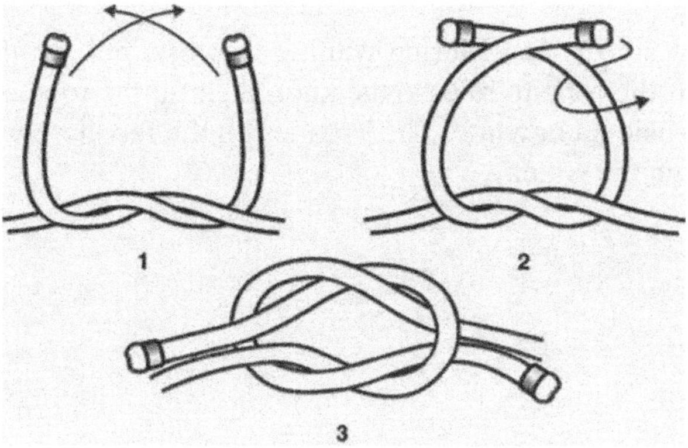

Bowline

The bowline forms a loop that will not tighten or slip under strain. It is easily untied.

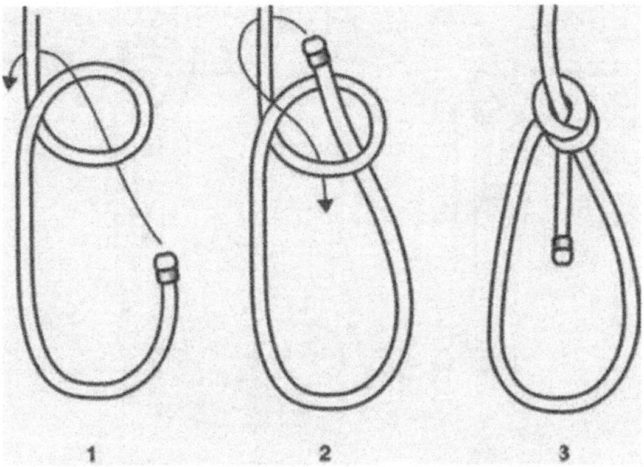

Hitches

Hitches are used to dress (prepare) knots and secure loose ends.

Half Hitch

The half hitch is used to tie a rope to a tree or to a larger rope. It will hold against a steady pull, but is not a secure hitch. It is frequently used to secure the free end of a rope.

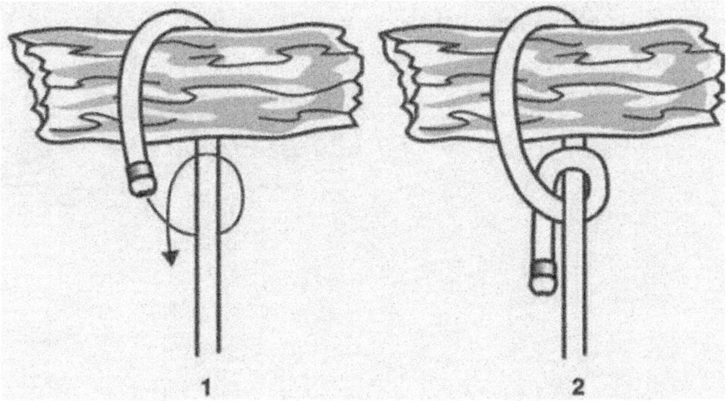

Two Half Hitches

Two half hitches can be used to secure the running end of a rope.

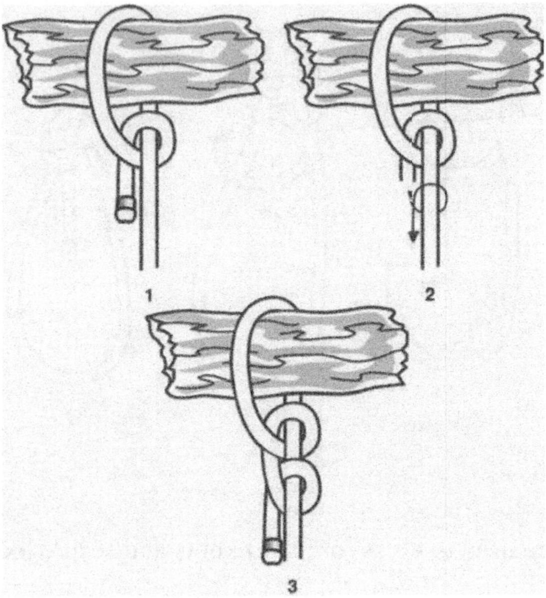

Round Turn and Two Half Hitches

A round turn and two half hitches can be used to fasten a rope to a tree. This hitch does not jam.

Butterfly Knot

The butterfly knot is used for anchor lines. Tied properly, it will not tighten on itself to the point that it cannot be easily untied. The butterfly knot can be used to pull a rope bridge taut. This knot can be used to tighten a fixed rope when mechanical means are not available. It will not jam if a stick is placed between the two upper loops.

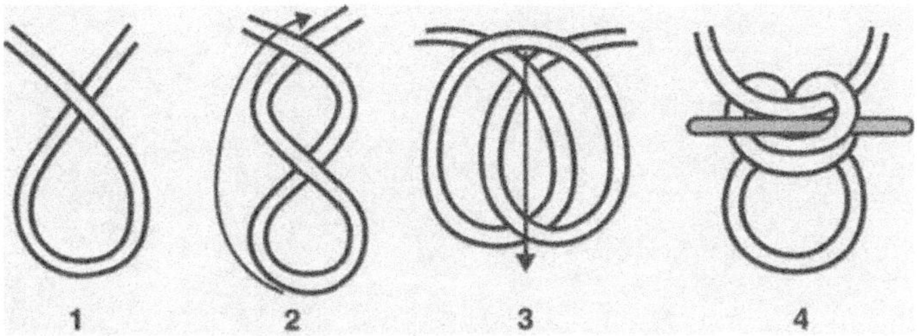

Table of Contents

Part I. Scouting

Chapter 1. Fundamentals of Scouting

1001 Purpose 1-1
1002 Required Scouting Skills 1-1

Chapter 2. Terrain, Maps, and Direction

2001 Terrain Features 2-1
2002 The Lensatic Compass 2-1
2003 Stars, Sun, and Other Features 2-5
2004 Range Determination 2-6

Chapter 3. Enemy Activity

3001 Estimating Enemy Strengths 3-1
3002 Interpreting Signs and Tracks 3-1
3003 Knowing the Enemy 3-2

Chapter 4. Daylight Scouting

4001 Cover and Concealment 4-1
4002 Camouflage 4-2
4003 Individual Movement 4-3
4004 Route Selection 4-5

Chapter 5. Night Scouting

5001 Night Vision 5-1
5002 Appearance of Objects 5-3
5003 Sounds 5-3
5004 Smells and Touch 5-3
5005 Clothing and Weapons 5-3
5006 Concealment 5-3
5007 Aids to Night Scouting 5-4
5008 Aids to Night Movement 5-4
5009 Locating and Plotting the Enemy at Night 5-6
50010 Routes of Movement 5-8

Chapter 6. Observing and Reporting

6001 Observation Posts 6-1
6002 Reporting 6-2

Chapter 7. Scouting Fire Team

7001 Positioning 7-1
7002 Locating Enemy Positions 7-2
7003 Action With an Attacking Platoon 7-2
7004 Action With an Enveloping Unit 7-5

PART 2. INFANTRY PATROLLING

Chapter 8. Fundamentals of Infantry Patrolling

8001 Definitions 8-1
8002 Relation of Patrolling to Scouting 8-1
8003 Purpose 8-1
8004 Types of Patrols 8-1
8005 Training 8-2
8006 Keys to Successful Patrolling 8-3

Chapter 9. Patrol Organization

9001 General Organization 9-1
9002 Task Organization 9-1

Chapter 10. Patrol Preparation

10001 Mission 10-1
10002 Factors Influencing Patrol Size 10-1
10003 Commander's Duties 10-1
10004 Patrol Leader Duties 10-2

Chapter 11. Movement to and Return from the Objective Area

11001 Passage of Lines 11-1
11002 Organization for Movement 11-1
11003 Control Measures for Movement 11-4
11004 Precautions at Danger Areas 11-5
11005 Hide 11-6
11006 Immediate Actions Upon Enemy Contact 11-6
11007 Patrol Leader's Action in a Developing Situation 11-10
11008 Return From Objective Area 11-10

Chapter 12. Reconnaissance Patrols

12001 General Missions 12-1
12002 Specific Missions 12-1
12003 Types of Reconnaissance 12-2
12004 Task Organization 12-2
12005 Size of Reconnaissance Patrols 12-3

12006 Reconnaissance Equipment 12-3
12007 Reconnaissance Patrol Actions at the Objective Area. 12-3

Chapter 13. Combat Patrols

13001 Task Organization 13-1
13002 Equipment 13-1
13003 Raid Patrols 13-1
13004 Contact Patrols. 13-3
13005 Ambush Patrols 13-3
13006 Security Patrols 13-7
13007 Urban Patrols 13-8

Chapter 14. Information and Reports

14001 Reporting 14-1
14002 Captured Items. 14-1
14003 Prisoners 14-2
14004 Patrol Report 14-2
14005 Patrol Critique 14-2

Appendix A. Patrol Warning Order A-1

Appendix B. Patrol Order B-1

Appendix C. Patrol Evaluation Checklist C-1

Appendix D. Ambush Formations D-1

Appendix E. Acronyms. E-1

Appendix F. References F-1

PART I. SCOUTING

CHAPTER 1. FUNDAMENTALS OF SCOUTING

Scouting involves observing terrain and/or the enemy, and accurately reporting those observations. Scouting requires proficiency in the use of weapons, cover and concealment, route selection, and skill in unobserved day or night movement.

1001. PURPOSE

When an infantry unit is not actively fighting the enemy, it should be actively searching for the enemy, which is the primary purpose of scouting. The unit attempts to keep the enemy off balance while making preparations for further attacks. Physically locating and keeping the enemy off balance are normally accomplished by small units ranging from a two-person scouting party to a squad-size patrol.

Infantrymen are sent out as scouts or as members of a patrol because the commander needs information about the enemy, terrain, and the location of friendly troops. The lives of the entire unit may depend upon the success or failure of a scout or patrol and the accuracy and timeliness of the report. The success of the scout or patrol will depend upon their training, preparation by the commander, and understanding their mission and the commander's requirements.

To wage combat successfully, a commander must have accurate, detailed, and timely information about the enemy, the terrain, and adjacent friendly units. Well-trained scouts and capably led patrols are among the most effective means the commander has for acquiring the information necessary to plan tactical actions and make decisions in execution.

1002. REQUIRED SCOUTING SKILLS

To be effective, a scout must be able to—

- Recognize terrain features.
- Read a map and determine direction.
- Practice and implement the principles of cover and concealment.
- Fully utilize movement and route selection.
- Know the enemy (estimate enemy unit composition and strength).
- Observe and report information accurately.
- Select routes and move through numerous types of terrain.

Chapter 2. Terrain, Maps, and Direction

A scout must understand map symbols, identify elevations from contour interval lines, scale distance on a map, relate natural and man-made features shown on the map to the actual features on the ground, plot a course from one point to another, and locate his current position. To relate a map to the actual terrain and its features, a scout must be able to orient it to the ground using a compass, two points, a watch, and the sun or the stars.

2001. TERRAIN FEATURES

Since the infantry works and fights on the ground, terrain ashore information that scouts gather and report on is of great importance. Hills, valleys, woods, and streams are the forms and growths commonly referred to as natural land features. Artificial or man-made features include houses, bridges, and railroads. Figure 2-1 shows some important terrain features.

2002. THE LENSATIC COMPASS

The best method of finding direction, during both day and night, is with a compass. The lensatic and M2 are the two types of compasses issued to infantry units. Both work on the same principle.

The standard compass for general use in the Marine Corps is the pivot-mounted lensatic compass, so called because azimuths are read through a magnifying lens in the eyepiece. Figure 2-2, on page 2-2, shows the lensatic compass and its nomenclature. The meter graphic scale on the side of the compass is 1:50,000, which is the most used scale in military mapping. The graphic scale is useful in the field as a straightedge, an aid in orienting the map, and a means of reading map grid coordinates. The plastic dial is graduated in both degrees and mils. Numbers on the dial are printed in black. There are luminous markings on the bezel, floating dial, and on both ends of the sighting wire, plus a 3-degree bezel serration and clicking device that permit reading azimuths at night. The compass is carried in a nylon case that may be attached to the cartridge belt.

Magnetic compasses are affected by the presence of iron or magnetic fields. Consequently, the scout should not be within the influence of local magnetic

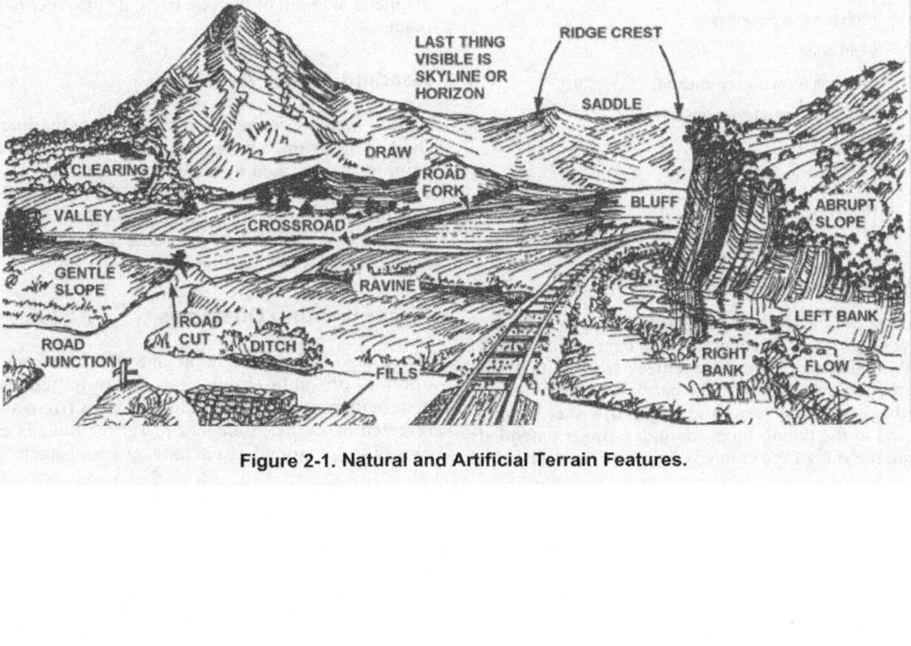

Figure 2-1. Natural and Artificial Terrain Features.

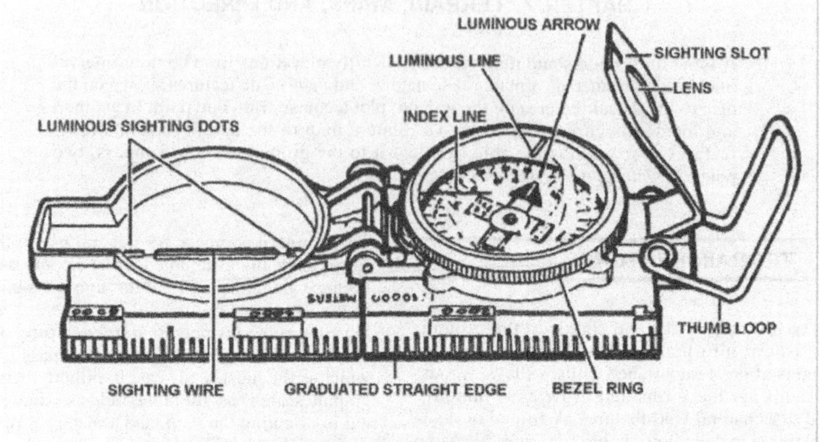

Figure 2-2. Lensatic Compass.

attraction while using a compass to determine direction. The rifle, pistol, and other metal objects must be laid aside when reading the compass. The minimum distances the scout should be from visible masses of iron and electrical fields of magnetism for the compass to provide accurate readings are provided below.

Magnetic Fields	Minimum Distances (meters)
High tension powerlines	60
Field guns	20
Vehicles (wheeled or tracked)	20
Telephone and telegraph wires	10
Barbed wire	10
Machine gun	3
Rifle, pistol	1

Reading an Azimuth

To read an azimuth to any point, the cover of the compass is raised to an angle of 90 degrees in relation to the index face, and the eyepiece is lifted to a 45-degree angle in relation to the bezel, or so the numbers on the dial can be seen. The thumb of either hand is placed in the thumb loop, the index finger extended along the side of the compass case, and the remainder of the hand closed. The closed hand and wrist are grasped with the other hand. The elbows are drawn in close to the body, forming a firm foundation for the compass. The eye is placed to the lens on the eyepiece. The compass is pointed at the object or point to which the azimuth is to be read. A sighting is taken through the sighting slot in the eyepiece, and the point is lined up with the sighting wire in the cover. The compass is held until the dial steadies; then the reading is taken through the lens of the eyepiece. This reading is the magnetic azimuth of the line from the observer to the point.

Reading a Back Azimuth

A back azimuth is the direction opposite the line of sight. If the azimuth is less than 180 degrees, the back azimuth is obtained by adding 180 degrees. If the azimuth is greater than 180 degrees, the back azimuth is obtained by subtracting 180 degrees. Back azimuths are used to determine a return route or to resection to determine a current position.

Circumventing Obstacles

When a scout is traveling on an azimuth and comes upon an obstacle—such as a contaminated area, minefield or swamp—the following steps (sometimes referred to as the *90-degree offset* method) is employed to go around or circumvent the obstacle and resume movement along the original azimuth (see fig. 2-3). The steps are as follows:

1. Move up to the obstacle and make a full 90-degree turn to the right (or left).

2. Walk beyond the obstacle, keeping track of the distance in paces or meters.

3. Stand at the end of the obstacle, face in the original direction of march, and follow that azimuth until the obstacle has been passed.

4. Make a 90-degree turn to the left (or right) and move the distance previously measured to return to the original line of march.

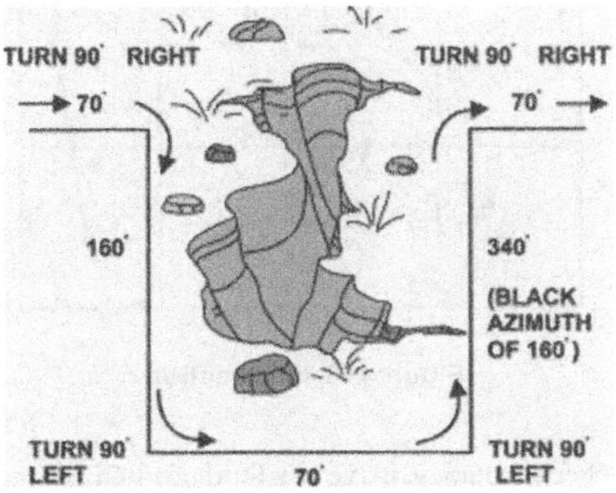

Figure 2-3. Circumventing an Obstacle.

Following an Azimuth During the Day

The eye is placed to the lens in the eyepiece and the compass moved until the desired azimuth reading is visible beneath the fixed index. Without moving the compass, the vision is shifted from the lens through the sighting slot in the eyepiece, and a sighting is taken out beyond the sighting wire in the cover. A prominent terrain feature on this line of sight is selected, the compass closed, and the landmark approached. When the scout reaches the landmark, the procedure is repeated.

Following an Azimuth at Night

It is necessary to prepare and set the compass before departing on a night movement because at night only the luminous parts of the compass can be seen. To prepare the compass for night use, the luminous parts must be fully charged by sunlight or artificial light, such as a flashlight. To set a compass—

1. Move the compass so that the desired azimuth on the dial is directly under the index line on the lower glass.

2. Rotate the upper movable glass so that the luminous line is directly above the north arrow of the dial.

3. Set the compass for marching at night on the specified azimuth.

—OR—

1. Face the general direction of movement.

2. Line up the north arrow and the luminous line on the bezel with the luminous sighting dots.

3. Hold the compass still with one hand and grip the knurled bezel ring with the other hand.

4. Turn the bezel ring the prescribed number of clicks in the proper direction, remembering that each click equals 3 degrees. For example, to set an azimuth of 21 degrees, the bezel ring would be turned seven clicks to the left.

5. Turn the whole compass until the north needle lines up with the luminous line. The compass is then set on the desired azimuth. The azimuth is the line formed by the two luminous sighting dots on the inside of the cover.

To march on a preset azimuth during night movement, open the compass and move it so the north arrow is directly below the luminous line (see fig. 2-4). Move

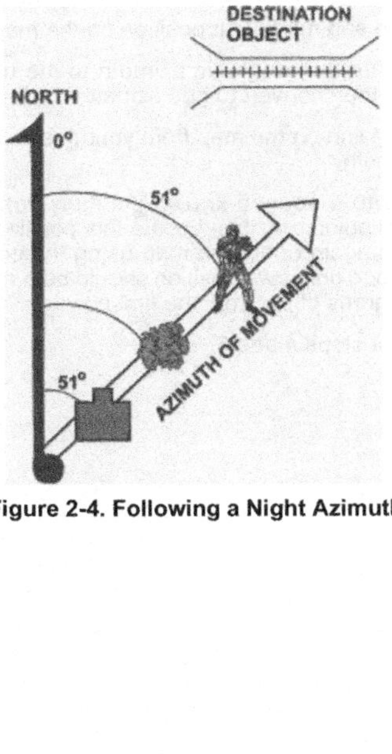

Figure 2-4. Following a Night Azimuth.

in the direction of the line formed by the two luminous sighting dots. It is necessary to refer to the compass more frequently at night than during the day. If stars are visible, find a prominent star along the azimuth of movement to use as a reference point. When the view of the sky is restricted by overcast conditions or vegetation, send a scout forward along the azimuth of movement to the limit of visibility. This scout is guided along the azimuth of movement by a stationary navigator. When the scout reaches the limit of visibility, the navigator moves to the scout's location. This process is repeated until the destination is reached.

A more rapid method for reaching the scout's destination is to equip the navigator with a compass. The navigator can set the compass as explained earlier and the scout proceeds providing security 180 degrees to the front on the specified azimuth, receiving right and left corrections from the navigator while both are on the move. The point scout must stay within visual range of the navigator. If available, a strip of white or luminous tape on the back of the point scout's helmet will assist.

Intersection

Intersection is the location of an unknown point by successively occupying at least two, preferably three known positions and sightings on the unknown point. It is used to locate features not depicted on the map or not readily identifiable. To determine an intersection, perform the following steps (see fig. 2-5):

1. Orient the map using the compass.

2. Locate and mark your position on the map.

3. Measure the magnetic azimuth to the unknown position; then convert to grid azimuth.

4. Draw a line on the map from your position on this grid azimuth.

5. Move to a second known position from which unknown point is visible. Locate this position on the map and again orient the map using the compass. The second unknown position should be a minimum of 30 degrees offset from the first position.

6. Repeat steps 4 and 5.

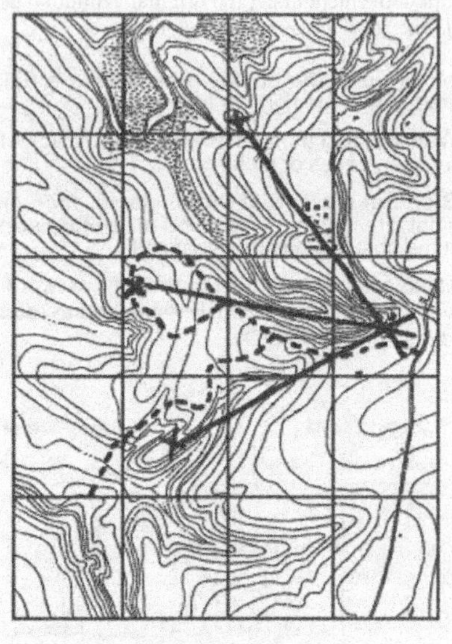

Figure 2-5. Intersection.

To check accuracy, move to a third position and repeat steps 1 thorough 4. Where the lines cross is the location of the unknown position. Using three lines, a triangle is sometimes formed—called the *triangle of error*—instead of an intersection. If the triangle is large, recheck your work to find the error. Do not assume that the position is at the center of the triangle.

Resection

Resection is the location of the user's unknown position by sighting on two or three known features that are identifiable on the map. To determine a resection, perform the following steps (see fig. 2-6):

1. Orient the map using the compass.

2. Locate two or three known positions on the ground and mark them on the map.

3. Measure the magnetic azimuth to a known position then convert to grid azimuth.

4. Change the grid azimuth to a back azimuth and draw a line on the map from the known position back toward the unknown position.

5. Repeat step 3 and step 4 to determine a second known position.

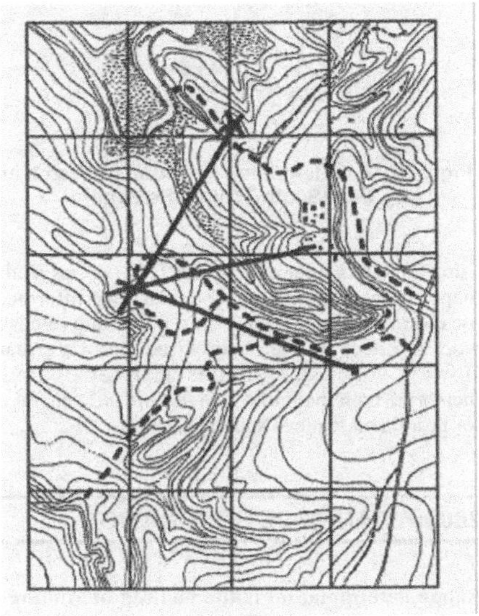

Figure 2-6. Resection.

To check accuracy, repeat the steps above for a third known position. The intersection of the lines is your location. Using three lines, a triangle of error may be formed. If the triangle is large, recheck.

2003. STARS, SUN, AND OTHER FEATURES

In rare cases when a scout is without a compass, the following examples are alternate means to determine direction. When using constellations to determine direction, identify your location's Temperate Zone. A Temperate Zone is the area between the tropics and the polar circles.

At night, the stars provide an excellent means of maintaining a line of march. In the North Temperate Zone (north of the equator), the Big Dipper constellation is one key to determining direction of true north. It is made up of seven fairly bright stars in the shape of a dipper with a long curved handle (see fig. 2-7). The two stars that form the side of the cup farthest from the handle, used as pointers, are situated in the direction of a bright star that is about five times the distance between the two stars of the dipper cup. This bright star is the North Star and is directly over the North Pole. The pointers always designate the North Star, which is the direction of true north.

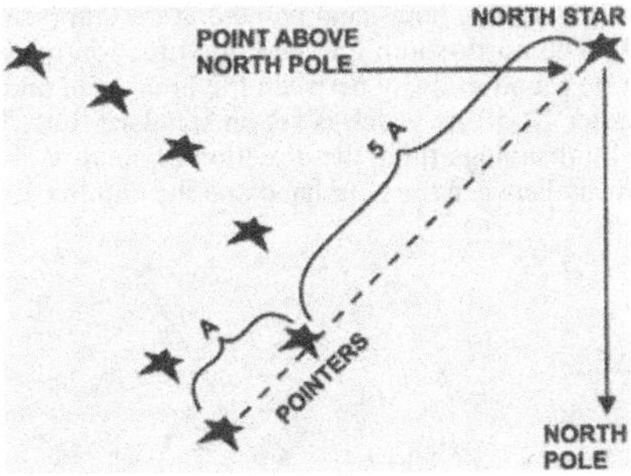

Figure 2-7. Locating the North Pole.

In the Southern Hemisphere, true south is determined in relation to the Southern Cross, a constellation composed of five stars. Two bright pointer stars in the vicinity of the Southern Cross serve as locators to help locate true south (see fig. 2-8). The outer four stars are

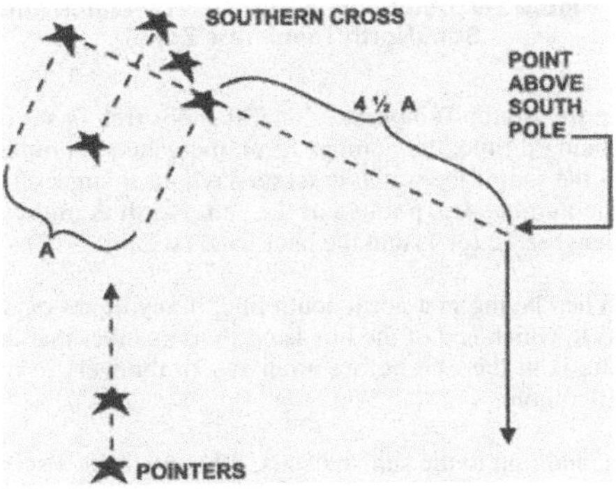

Figure 2-8. Locating the South Pole.

fairly bright and form a cross. This cross is imagined as the frame of a kite. A straight tail, four and one half times as long as the length of the kite itself, is put on the kite using finger widths for a measuring stick. The end of this tail will be close to a position directly over the South Pole. Usually, it will not be possible to see a star in the immediate vicinity, because there is no bright star visible directly above the South Pole.

During daylight hours, a watch and the sun can be used to determine direction within 8 degrees. In the North Temperate Zone, the watch is held horizontally, face up, and the hour hand pointed at the sun (see fig. 2-9). The north-south line and the direction of south can be found midway between the hour hand and the number 12, if the watch is set on standard time. If in daylight savings time, the direction of south is found midway between the hour hand and the number 1.

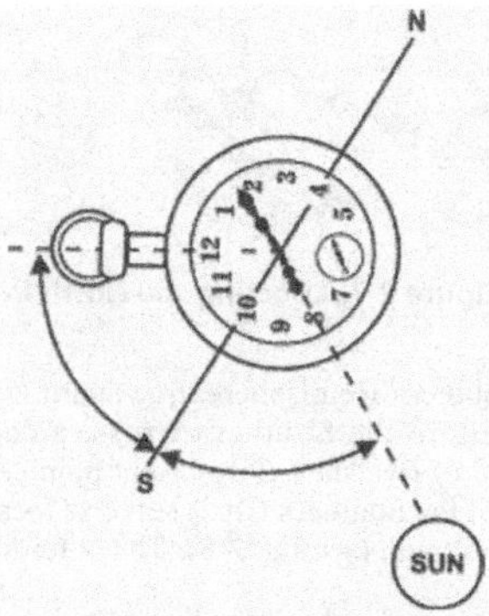

Figure 2-9. Determining Direction by Watch and Sun (North Temperate Zone).

In the South Temperate Zone, if the watch is set on standard time, the number 12 on the watch is pointed at the sun; if the watch is set on daylight savings time, the number 1 is pointed at the sun. North is midway between 12 (or 1) and the hour hand (see fig. 2-10).

When laying in a north-south line, if any doubt exists as to which end of the line is north, remember that the sun is in the east before noon and in the west in the afternoon.

In addition to the sun and stars, other methods a scout without a compass can use to determine direction include determining prevailing wind direction and

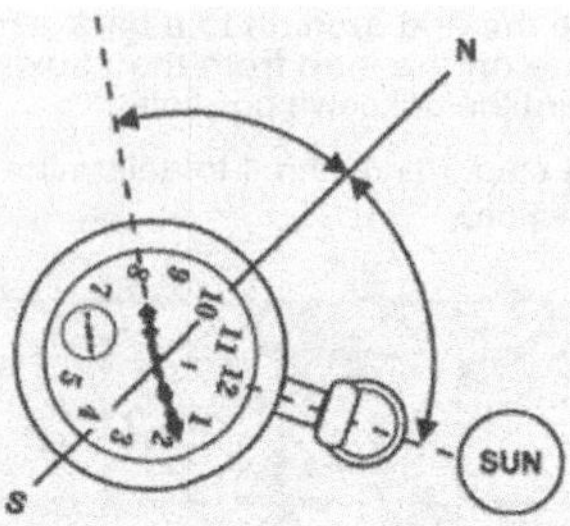

Figure 2-10. Determining Direction by Watch and Sun (South Temperate Zone).

using a mountain for orientation. By previous study of maps and photographs, a scout can keep informed of location and direction by using a distinctive edge of woods, a deep ravine or the direction of a stream's flow. A scout should constantly evaluate and memorize both the immediate terrain and general area for prominent features and landmarks.

2004. RANGE DETERMINATION

Range determination is the method of finding the distance between an observer and an enemy target or an object. By accurate range determination, the members of a given unit can set their sights correctly and place effective fire on enemy targets. The degree of accuracy is dependent on several factors, such as terrain relief, time available, and experience of the observer.

Mental Estimation

A mental distance estimate is made using a known unit of measure. Distance is estimated to the nearest 100 meters by determining the number of known units of measure between the observer's position and a target. For example, a football field, which is 100 yards, can be used as a known unit of measure for determining the distance between an observer's position and a target. For longer distances, progressive estimation may be necessary. To do this, the observer determines the number of units of measure to an intermediate point and doubles the value. The observer should consider the effects in table 2-1 in estimating distances.

Table 2-1. Effects to Consider in Mental Estimation of Distances.

Objects Appear Nearer	Objects Appear More Distant
In bright light.	In poor light or in fog.
In clear air at high altitude.	Only a small part of the object can be seen.
The background is in contrast with the color of the object.	The background is similar in color to that of the object.
The observer is looking down from a height.	The observer is looking over a depression, most of which is visible.
The observer is looking over a depression, most of which is hidden.	The observer is kneeling or sitting, especially on a hot day, when the ground is moist.
The observer is looking down a straight feature such as a road.	
The observer is looking over water, snow, or a uniform surface such as a cultivated field or desert.	

Estimating in Good Visibility

When visibility is good, distances can be estimated by using the appearance of tree trunks, branches, and foliage (as seen by the naked eye) in comparison with map data. Table 2-2 is a guide for wooded terrain. Table 2-3 is a guide for urban environments.

Estimating From a Terrain Study

The Marine should always use terrain/map analysis to assist in estimating distances. When the Marine is looking in a specific direction, the estimation of distance can be enhanced by studying the terrain and comparing it with the map. Particular emphasis should be given to color contrasts of terrain features seen along the observer-target line (OTL). For example, the distance across successive ridge lines or depressions in the distance may be identifiable by only slight changes of color to the eye. Different colors of grass might reveal a hidden terrain feature such as a stream.

Table 2-2. Estimating Distance in Wooded Terrain.

Distance in Meters	Tree Description
1,000	Trunk and main branches are visible. Foliage appears in cluster-like shape. Daylight may be seen through the foliage.
2,000	Trunk visible, main branches distinguishable, foliage appears as smooth surface. Outline of foliage of separate trees distinguishable.
3,000	Lower half of trunk visible. Branches blend with foliage. Foliage blends with adjoining trees.
4,000	Trunk and branches blend with foliage and appears as a continuous cluster, smooth in appearance. Movement of foliage due to wind cannot be deleted.
5,000 and beyond	Whole area covered by trees and appears smooth and dark.

Table 2-3. Estimating Distance in Urban Terrain.

Distance in Meters	Object Identified by the Unaided Eye
1,000	Lone tree trunk
1,500	Individuals and horsemen
3,000	Chimneys on rooftops
4,000	Windows in houses
4,000–5,000	Individual houses in populated area
8,000–9,000	Villages and individual houses
15,000–18,000	Large houses, towers, and steeples

Chapter 3. Enemy Activity

A commander often acts on information furnished by scouts. Therefore, scouts must aim at absolute accuracy in reporting enemy activity. This chapter discusses estimating enemy strengths, interpreting signs and tracks, and knowing the enemy.

3001. ESTIMATING ENEMY STRENGTHS

If troops cannot be counted, their strength may be estimated by: noting the length of time it takes various types of moving columns to pass given point, the area required of a unit in camp or bivouac, or the front on which they are deployed. When the ground is dry, infantry on the march raise a low, thick cloud of dust, and motor vehicles or mechanized units raise a thick, rapidly moving cloud. Additionally, through practice, a scout may gain information as to the strength and composition of enemy forces by listening to noises and observing lights, fires, and smoke. A scout gains valuable experience in estimating enemy strengths by observing friendly forces in camp, on the march, and deployed. The knowledge scouts gain during field exercises of the appearance and tactical dispositions of squads, platoons, companies, and larger units will be of great assistance in estimating the strength and composition of enemy units observed under various conditions.

3002. INTERPRETING SIGNS AND TRACKS

In addition to estimates made through direct observation, a scout may often be able to estimate size, composition, direction, rate of movement, condition, discipline, state of training, and morale of enemy forces through signs and tracks left behind.

Signs

The examination of vacated enemy positions provides valuable information. The size of a bivouac or defense area ordinarily indicates the number of enemy occupants. Clothing, ration containers, dumps, etc., further indicate the quantity of the departed enemy force. The condition of the bivouac area and amount of material abandoned give an indication of the enemy morale, training, and discipline. A well-policed area indicates good discipline. Rubbish, ration and smoking residue, and nonessential personal items of equipment adrift indicate a lower state of morale, training, and discipline. Stores and material left behind in good condition may indicate a hasty movement or withdrawal. Burned or destroyed materials indicate a deliberate, orderly withdrawal or movement. Letters, insignia, and other articles may reveal the identity of the enemy unit.

In the case of a moving enemy, the distance between periodic halts indicates the rate of march if enemy habits relative to marches and halts are known. Condition of the halt areas indicates the state of morale, training, and discipline.

The physical condition of enemy dead and wounded and their personal equipment and weapons are reported. The general condition and state of maintenance of destroyed or abandoned vehicles should also be reported.

Tracks

A track is a mark left on the ground by the passage of a person or object. Examination of tracks reveals information about the enemy.

Troops

A few tracks overlapping each other on both sides of a road or trail may indicate a patrol in staggered formation. A large number of tracks indicates troops in column formation. A large column will wear a dry road smooth and flat. In damp terrain, a freshly made track will have sharp edges; ordinarily, signs of moisture will disappear in about 15 minutes. A runner's toes are dug into the ground; a walker's footprint is fairly even.

Vehicles

The type of track indicates whether the vehicle is wheeled or tracked. A scout acquires the necessary experience to make the proper determination by observing vehicle tracks during training.

The direction of travel can be determined by the way tracks pass across ruts, by impressions on the edges of holes in the ground, how water is splashed from puddles, or by the way grass, twigs, and branches are bent; for example—

- A vehicle (wheeled or tracked) entering a rut pushes dirt into the rut and leaves an indentation on the exit side of the rut.
- A wheel going over holes in the ground leaves a deeper impression on the edge toward the direction of travel.
- The side of a puddle with the greater splash indicates the direction of travel of the vehicle.
- When traveling cross-country, the direction in which grass is bent and/or twigs, branches, and bushes are broken indicates direction of travel.

A general rate of speed can be estimated by the amount of water or mud splattered. A fast-moving vehicle will throw larger amounts of water or mud a greater distance to the front and sides than a slow-moving vehicle, and it leaves a deeper impression on the exit edges of holes. The faster the travel, the deeper the impression.

3003. KNOWING THE ENEMY

A scout should learn as much as possible about enemy psychology, habits, organization, and tactics. The more knowledge gained about the enemy, particularly the enemy's normal security measures, the better the scout's chances are to observe and obtain accurate information with minimum risk to the mission's success. Scouts gain much of this knowledge through experience, but they also gain a great deal of their preliminary information, particularly that pertaining to enemy organization and tactics, during training and may be updated by unit commanders and intelligence officers.

CHAPTER 4. DAYLIGHT SCOUTING

A scout must be able to operate in all types of terrain and under all conditions of visibility. He must be thoroughly familiar with the principles for using cover and concealment, camouflage, individual movement, and route selection, both to and from the objective.

4001. COVER AND CONCEALMENT

Cover is protection from the fire of hostile weapons. Concealment is protection from observation or surveillance from hostile air and ground observation, but not from hostile fire. Both cover and concealment are divided into two main categories: natural and artificial. Natural cover includes small hills, ditches, rocks or vegetation. Fighting holes, bunkers, and brick walls are examples of artificial cover. Some features, such as buildings, provide both cover and concealment. In deciding whether to seek cover or concealment, a scout must make the best choice to complete the mission (see fig. 4-1).

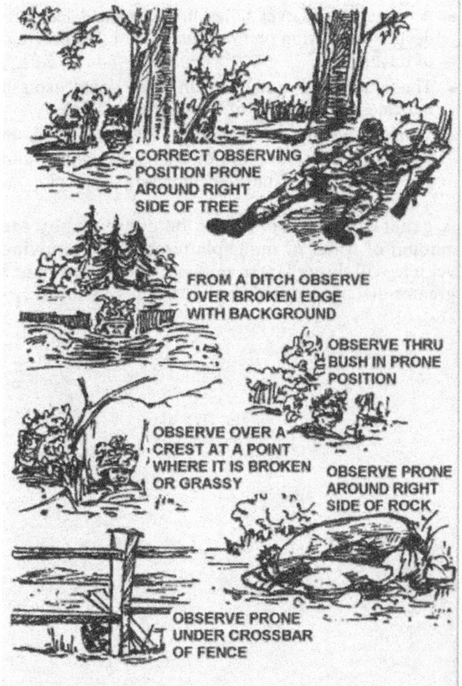

Figure 4-1. Correct Use of Cover.

Concealment Principles

Concealment principles are as follows:

- Remain motionless while observing. Anything in motion attracts the eye.
- Use all available concealment.
- Observe from the prone position (it offers a low silhouette and makes detection by the enemy difficult.
- Expose nothing that reflects light.
- Blend with the background because contrasting colors are noticeable.
- Remain in the shade because moving shadows attract attention.
- Distort or change the regular outline of objects. Most military objects have distinctive shapes that make obvious shadows and silhouettes.
- Avoid the skyline. Figures on the skyline can be seen from great distances and are easily identified by their outlines.

Concealment Techniques

Concealment techniques are as follows:

- When observing, the scout looks around an object's side (unless it is transparent) and prepares to fire, if necessary, around the side of or, if possible, through an object.
- Looking or firing over an object can make the scout an easily visible target for the enemy. If the scout must fire over the top of concealment or cover, the outline of the head or helmet should be broken or distorted.
- Upon the approach of an airplane, the scout takes a prone position, turns face-down, and remains motionless. If surprised by an airplane, the scout remains in place and does not look up.

- The scout covers exposed body parts such as the face, back of the neck, and hands with grease paint, mud or other materials to reduce sun reflection.
- Camouflage for equipment can be improvised from garnishing or sandbags to prevent sun reflection.
- In snowy terrain, white overgarments are worn.
- The helmet cover outline should be distorted.

4002. CAMOUFLAGE

Camouflage is the use of concealment and disguise to minimize the possibility of detection and/or identification of troops, material, equipment, and installations. The purpose of camouflage is to provide concealment of military objects from enemy observation. Camouflage is also used to conceal an object by making it look like something else. A scout's mission usually requires individual and equipment camouflage. If natural camouflage is not adequate, the position is camouflaged. In using camouflage, remember that objects are identified by their form (outline), shadow, texture, and color. The principal purpose of camouflage in the field is to prevent direct observation and recognition.

Individual Camouflage

Successful individual camouflage involves the ability to recognize and take advantage of all forms of natural and artificial concealment available (vegetation, soil, debris, etc.) and knowledge of the proper use of artificial camouflage materials.

Aids to Individual Camouflage

A scout must recognize the terrain's dominant color and pattern and must change the appearance of clothing and equipment accordingly in order to blend and not contrast with the terrain (see fig. 4-2).

The helmet is camouflaged by breaking up its shape, smooth surface, and shadow. Use of a helmet cover works best. In the absence of a helmet cover, mud can be irregularly blotched on the helmet to disguise its form and dull the surface. A helmet cover may be improvised from irregularly colored cloth or burlap to blend with the background. Foliage can be draped to prevent the visor of the helmet from casting a dark shadow across the face. Foliage should not stick up like plumes because any head movement will give away the position.

A small, thin bush in the shadow of a large bush makes a good observation point. Lone trees, rocks, fence corners, and outstanding landmarks are easily picked up by the enemy as obvious observation posts.

If camouflage clothing is not available, other available clothing can be attached in irregular splotches of appropriate colors.

Exposed skin reflects light and attracts the enemy's attention. Even very dark skin will reflect light because of its natural oil. The buddy system is recom-mended when applying camouflage. Standard Marine Corps issue camouflage face paint sticks are two toned:

- Loam and light green for light-skinned troops, in all but snow regions.
- Sand and light green for dark-skinned troops.
- Loam and white for troops in snow-covered terrain.

Shiny areas (forehead, cheekbones, nose, and chin) are painted with a dark color. Shadow areas (around the eyes, under the nose, and under the chin) are painted

Figure 4-2. Avoid Contrasting Backgrounds.

with a light color. Skin that is exposed on the back of the neck and hands is painted with a two-color combination in an irregular pattern (see fig. 4-3).

When standard issue face paint sticks are not available, burnt cork, charcoal or lamp black can be used to tone down exposed areas of skin.

Mud is used only in an emergency because it changes color as it dries and may peel off, leaving the skin exposed. Since mud may contain harmful bacteria, mud should be washed off as soon as possible.

Any equipment that reflects light should be covered with a nonreflective material that aids in the concealment of the weapon (for example, black electrical tape or mud). The straight line of the rifle or other infantry weapons may be very conspicuous to an enemy observer. The barrel and hand guard should be wrapped with strips of contrasting colored cloth or tape to break the regular outline. Mud or dirt dulls the reflecting surface of the stock, barrel, and bayonet where coloring has been worn. Lamp black may also be used on metal parts. The function of the weapon must not impaired.

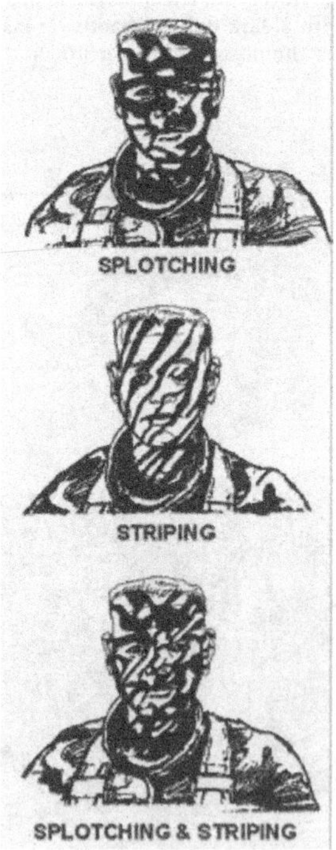

Figure 4-3. Face Camouflage.

If time, material, and surroundings permit, a ghillie suit should be constructed. (Refer to MCWP 3-15.3 *Scout Sniping*.)

Aids to Camouflage a Position

To successfully camouflage a position, the scout must remember to—

- Camouflage the position as soon as it is occupied.
- Avoid using too much material for camouflage. Even though natural materials are used, too much may make the object and its shadow stand out from its surroundings, thus attracting the attention of a hostile observer.
- Inspect completed camouflage work from the enemy's point of view to check effectiveness.

Continuous Camouflage

Camouflage around and on the scout's position must be maintained in a fresh condition as wilted and dead foliage can give the position away. If the mission dictates that the position should be occupied for longer periods, wilted foliage should be replaced during periods of reduced visibility.

4003. INDIVIDUAL MOVEMENT

Principles

The principles of individual movement are as follows:

- Scouts move from one concealed position to another. When not changing positions, they remain motionless.
- The scout's head is lifted slowly but steadily, without abrupt movements, to search for a new position.
- Scouts select the next stopping place before moving and ensures it is not contained by the enemy.
- Scouts change position on the run: springs up, runs with the body bent low, zigzags, quickly drops to the ground slightly to the right or left of the objective, then rolls or crawls to the desired position.

(Remember the phrase, "I'm up—he sees me—I'm down.")

Rushing

When starting from the prone position—

- Raise the head slowly and steadily and select a new position.
- Lower the head slowly, draw arms inward, cock right leg forward, and prepare to rush.
- Use one movement to raise the body by straightening both arms.
- Spring to your feet, stepping off with the left foot.
- Bend forward as low as possible when running. Never advance directly to the next position; always zigzag.

When hitting the deck—

- Stop.
- Plant both feet in place.
- Drop quickly to the knees and slide the hand to the heel of the rifle.
- Fall forward, breaking your fall with the butt of the rifle. (To confuse the enemy, roll over after hitting the deck and roll into firing position with feet, knees, and stomach flat on the ground.)
- Keep head down if you do not intend to fire.

When rolling over—

- Hit the deck and assume the prone position.
- Bring the rifle in close to the body, placing the rifle butt in the crotch.
- Roll over swiftly to confuse any enemy observers as to final intended location. Never reappear at the same place you went down.

Low Crawl

The low crawl is used when—

- Cover and concealment are scarce.
- The enemy has good observation over the area in which the scout is moving.
- Speed is not essential.

To perform the low crawl, keep the body as flat as possible against the ground. Grasp the rifle sling at the upper sling swivel. Let the balance of the rifle rest on the forearm and let the butt of the rifle drag on the ground. Keep the muzzle off the ground.

To start forward, push arms forward and pull right leg forward. To move forward, pull with arms and push with right leg. Change the pushing leg frequently to avoid fatigue.

High Crawl

The high crawl is used when—

- Cover and/or concealment are available.
- Poor visibility reduces enemy observation
- Greater speed of movement is required.

To perform the high crawl, keep body off the ground. Rest weight on forearms and lower legs. Cradle rifle in arms, keeping the muzzle off the ground. Keep knees well behind the buttocks to stay low.

Move forward, alternately advancing right forearm and left knee; then left forearm and right knee.

Movement Aids

Aids to movement include—

- Carrying only necessities. Additional weight causes premature fatigue and impedes free movement.
- Not disturbing birds or animals whose flight would betray your presence. If birds or animals are alerted, remain motionless under cover for a few minutes, as the enemy's attention may also be attracted.
- Moving during an incident that diverts attention, such as an airplane flight, a distant disturbance or sudden bursts of fire.
- Fog, smoke, or even light haze offer concealment for movement; however, the enemy may have thermoimagery and night vision devices. Therefore, darkness and smoke cannot be used as easily.
- Following a stream or road by staying as far away from them as possible while still keeping them in sight. Keep close to the dune line when moving along a beach.
- When moving through tall grass or similar growth, move when the wind blows, changing direction frequently. A straight route will be more readily noticed.
- Whenever possible, avoid areas of soft ground so as not to leave tracks.

- When crossing a road or water obstacle, choose crossing sites where the enemy's observation is restricted (an area in shadows or near a bend) and cross rapidly.

4004. ROUTE SELECTION

Prior to Movement

A scout and the immediate commander conduct a map reconnaissance before starting on a mission. This assists them in selecting the route according to available cover and concealment and any indicated enemy activity.

Prior to and during the course of the mission, move to an observation point to visually reconnoiter the terrain for movement and select the tentative route. It may be necessary to make wide detours around open spaces or those containing enemy patrols or other enemy activity.

Carefully study the country to be traversed and pay close attention to the general features, streams, ridges to be crossed, and their relation to the general direction to be taken (see fig. 4-4).

Make notes of terrain features and landmarks along the proposed route and rely on notes for guidance (see fig. 4-5 on page 4-6). Additionally, determine the compass direction and readings for each change of direction at the start. Finally, learn the location of unit boundaries and observation/listening posts as well as general location of other friendly or scouting parties. Be sure to avoid man-made and natural obstacles as they will slow progress and overall success of the mission. If possible, use the local populous as a source of intelligence. When returning to friendly lines, avoid using the same route.

En Route

En route, the actual advance will be a series of movements from one observation point to the next. The distance and route will depend on cover and terrain. Assess the cover, terrain, and any enemy or civilian activity to determine whether or not to modify the approach or return routes. Unless the mission requires it, avoid danger areas (for example, houses, villages, potential assembly or bivouac areas, roads, and streams) that may give away your position by being observed by the enemy. When required to reconnoiter danger areas, choose a covered approach and return, and make entry or passage as quietly and quickly as possible. If part of a larger effort, the approach and return should be covered by observation and fires of the other members of the scouting party or patrol.

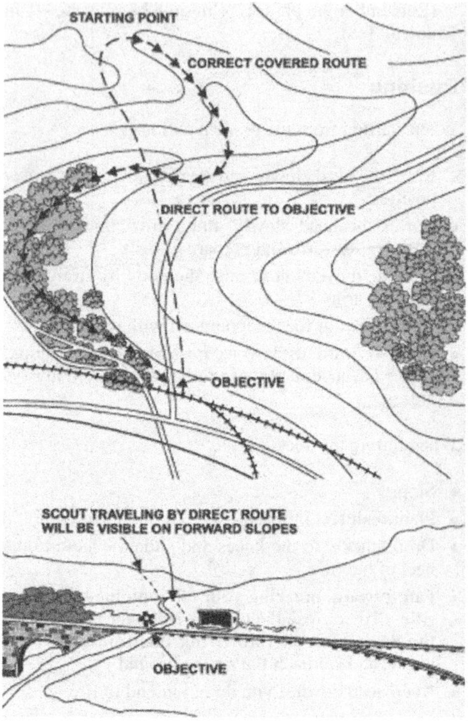

Figure 4-4. Choosing a Concealed Route of Advance from a Map.

Stream Crossings

When the crossing does not appear to be held by the enemy, advance upon it rapidly. If there are two or more scouts, one crosses while the other(s) provide protection. Note the length, width, depth, and approaches to a crossing. Observe the condition of the road or trail that crosses the stream, and report on the suitability of the crossing for use by tracked and wheeled vehicles. If the crossing is under observation by enemy, seek another crossing site or dash across to avoid detection.

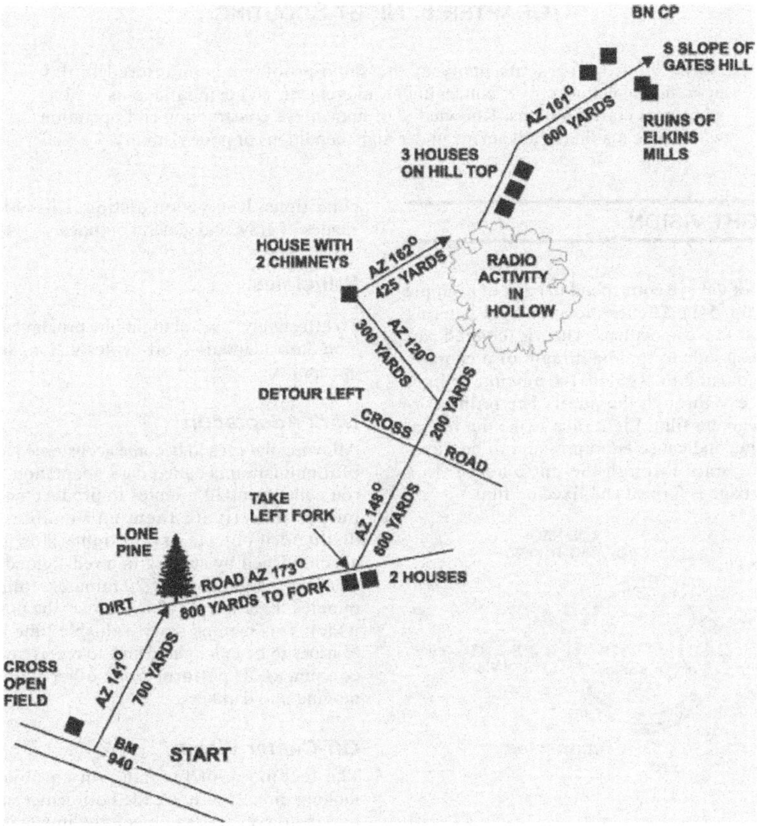

Figure 4-5. Proposed Route Sketch.

CHAPTER 5. NIGHT SCOUTING

Night scouting presents many of the same problems encountered in day operations-such as cover, concealment, movement, and camouflage-as well as additional considerations. Knowledge of human eye construction and operation will enable maximum advantage under night conditions or poor visibility.

5001. NIGHT VISION

Certain parts of the eye correspond to parts of a simple camera (see fig. 5-1). The lens focuses light entering the eye similar to a camera lens. The iris (colored part of eye) corresponds to the diaphragm of a camera, opening and closing to regulate the amount of light entering the eye through the pupil. The retina corresponds to camera film. Light rays strike the retina, form an image, and cause an impression to be transmitted to the brain through the optic nerve. In a camera, the image is formed and fixed on film.

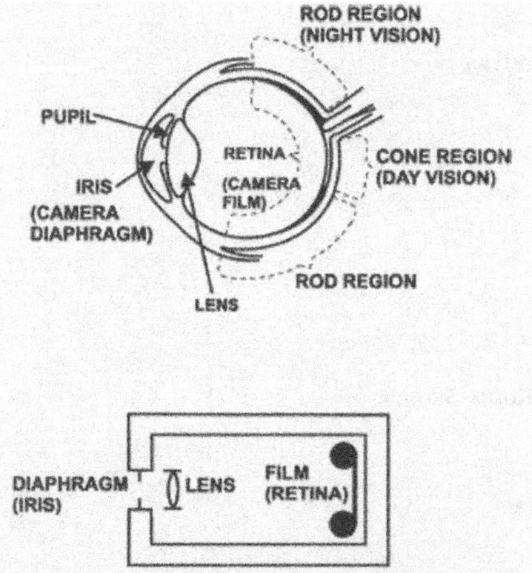

Figure 5-1. The Eye is Like a Camera.

The retina is composed of cone cells and rod cells, so-called because of their shapes. Cone cells distinguish color, shape, and sharp contrast. Because they are activated by light conditions, they are blind during periods of low illumination. Rod cells produce a chemical substance called visual purple that makes them active in darkness, low illumination or night conditions. Rod vision distinguishes black, white, shades of gray, and general outlines.

Principles

To effectively "see" at night, the principles of night vision dark adaptation, off-center vision, and scanning are applied.

Dark Adaptation

Allowing the eyes to become accustomed to low levels of illumination is called dark adaptation. It takes the rod cells about 30 minutes to produce enough visual purple to activate them and enable the eye to distinguish objects in dim light. This may also be accomplished by staying in a red-lighted area, or by wearing red goggles for 20 minutes, followed by 10 minutes in darkness (which allows the pupils to open wide). This method saves valuable time by allowing Marines to be in a lighted area to receive orders, check equipment, or perform some other function before moving into darkness.

Off-Center Vision

The technique of focusing on an object without looking directly at it is called off-center vision. When looking directly at an object, the image is formed on the cone region, which is not sensitive at night (see fig. 5-2 on page 5-2). When looking slightly to the left, right, above or below an object, the image is formed on the area of the retina containing rod cells, which are sensitive in darkness. The most sensitive area varies in individuals, but is usually found by looking 6 to 10 degrees away from an object; in effect, out of the corner of the eye (see fig. 5-3 on page 5-2).

Scanning

Off-center vision used to observe an area or an object is called scanning. When using rod vision, the visual purple in the rod cells bleaches or blacks out in 4 to 10 seconds and the object observed disappears. As the visual purple in the rod cells in one area bleaches out, the eyes must slightly shift to use fresh rod cells. Eyes

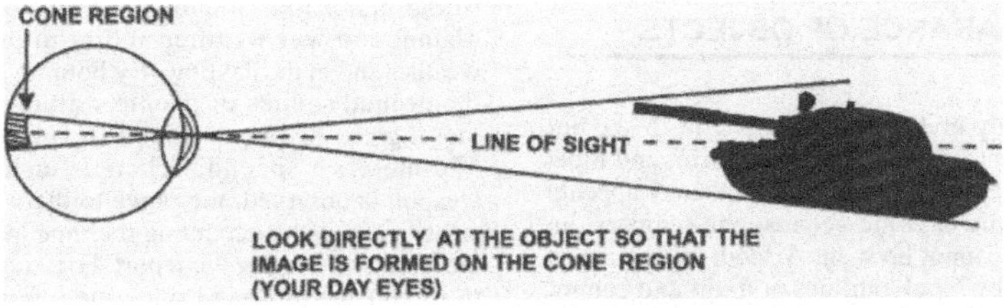

Figure 5-2. Day Vision.

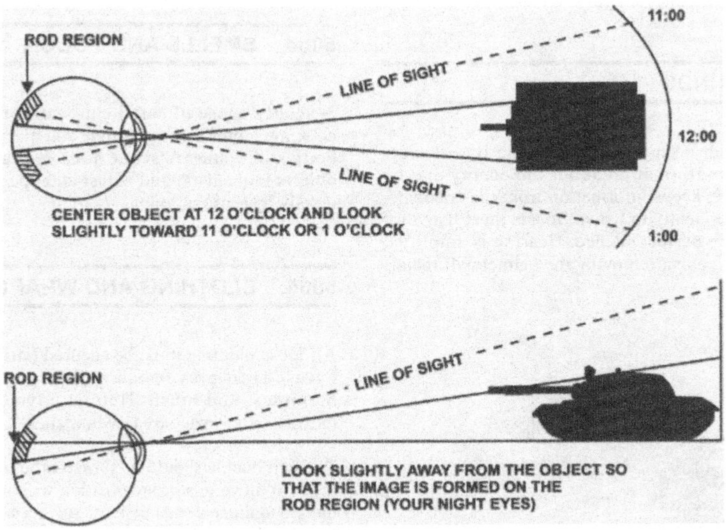

Figure 5-3. Night Vision.

should be moved in short, abrupt, irregular movements over and around the target (see fig. 5-4).

Preserving Night Vision

Night vision is quickly destroyed if bright light is allowed to enter the eye. When entering a lighted area or when observing in a temporarily lighted area (illumination, flares), one eye should be closed and covered to preserve its night vision. When the light goes off, fades or the lighted area is exited, the night vision retained by the protected eye enables it to see until the other eye adapts to the darkness. Red light helps preserve night vision, but like white light, it can be observed at long distances.

Factors that decrease night visual acuity include fatigue, lack of oxygen, long exposure to sunlight, alcohol, nicotine (within the past 48 hours), and age. When night vision has been attained, straining will not improve effectiveness; however, practice identifying objects at night will improve perception.

5002. APPEARANCE OF OBJECTS

Darkness not only makes objects difficult to see but also changes their appearance, distorts size, and blots out details. A tree visible against the night sky appears smaller than in the daytime because the twigs at the end of branches cannot be seen. A scout must train to identify objects by block outlines at night and cannot rely on details visible in daylight. Binoculars enlarge objects or parts of objects otherwise too small to be seen and help identify objects already spotted. Night observation devices increase night visibility and should be used whenever possible.

5003. SOUNDS

At night, sounds become very important. By listening, a scout gains information about the enemy and by exercising care, keeps information from the enemy. A scout stops frequently to listen. Scouts must listen for long periods in perfect silence. Hearing is amplified with the mouth open; removing the helmet will reduce sound distortion. Sounds are transmitted a greater distance in wet weather and at night than in dry weather and in the daytime. By holding the ear close to the ground sounds of people walking and vehicles moving can be heard. Sound travels approximately 370 meters a second. When a flash from a fired weapon is observed, the range to the weapons can be easily estimated by counting the time interval between the flash and hearing the report. For example, counting to three (one thousand one, one thousand two, one thousand three), indicates the distance is 1,110 meters. The cadence is determined by actual practice at known ranges.

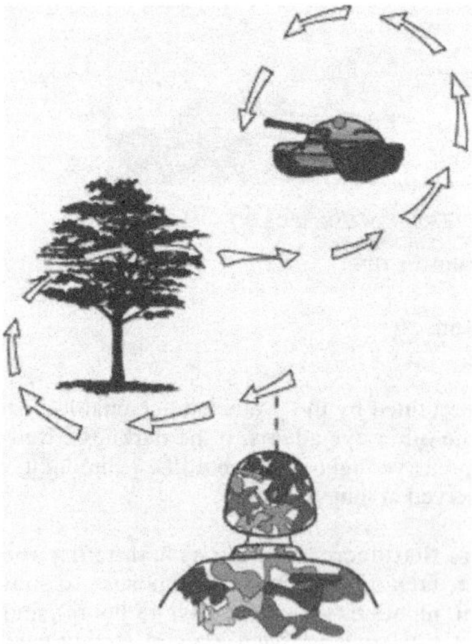

Figure 5-4. Scanning

5004. SMELLS AND TOUCH

A scout's sense of smell can warn of enemy fires, cooking, motor parks, gasoline and diesel engines, and bodies of water. A scout must feel and recognize objects in the dark, and adjust and operate equipment quietly by sense of touch.

5005. CLOTHING AND WEAPONS

All loose clothing must be secured (string or tape can be used) to prevent snagging on barbed wire, brambles, and brush. Helmet covers are worn to muffle sounds made by low branches.

The belt buckle should be turned around to the side in order to move in a prone position without scraping the buckle against stones or hard surfaces. Identification tags can be taped together to prevent rattling. Hands, face, and neck can be blackened so skin does not reflect light or appear as white spots in the darkness. (Refer to para. 4002.)

Scouts are normally armed with rifles. Rifle slings should be taped to prevent rattling. All weapons parts should be checked for glare elimination measures.

5006. CONCEALMENT

Although total darkness provides concealment, scouts must observe the same principles of concealment during moonlight conditions as in the daytime. Scouts should assume enemy employment of night observation devices and observe the principles of night movement such that presence will not be disclosed by noise when close to the enemy.

5007. AIDS TO NIGHT SCOUTING

Aids to night scouting include the following:

- Carry out scouting missions close to or within hostile positions on dark or rainy nights.
- Stifle a threatening sneeze by pressing fingers upward against the nostrils.
- Stifle a threatening cough by applying slight pressure with the finger on the Adam's apple.
- Stop a ringing sound that interferes with hearing by yawning.
- Speak softly rather than whispering if voice communication is necessary.
- Move boldly and rapidly when taking advantage of any sound-such as shelling, rustling wind or distant firing-to push forward if firing is taking place.
- Avoid shell craters and depressions in damp weather conditions if the enemy has employed chemical munitions.
- Move the eyes constantly; concentrating on one object too long will strain them.
- Take notice of the enemy's use of flares. When the enemy employs flares, few enemy patrols are apt to be out; when flares are not employed, the enemy's patrols are likely to be numerous.
- Drop to a prone position upon hearing a flare being fired and before it illuminates. Remain motionless while it is burning. If open or moving when a flare bursts in the air, freeze or drop quickly in the split second after the flare illuminates while the enemy is blinded. You are an easy target for the enemy if the flare bursts in the air or on the ground behind you. Never look at a flare. If you activate a trip flare, drop to the ground and crawl away from the illuminated area.
- Consider all patrols or individuals encountered as hostile until proven friendly. If encountering someone, crouch low, silhouetting the approaching person against the sky. At the same time, make yourself an indistinct target in case the person encountered is an enemy.
- Return fire only to avoid capture if fired on when close to enemy positions.

5008. AIDS TO NIGHT MOVEMENT

- Aids to night movement include the following:
- Move silently.
- Advance in stealthy legs. Each leg should follow some terrain feature that serves as a guide. When there are no terrain features to serve as guides, move in a straight or nearly straight line from one defined point to another, or maintain direction by using a compass.
- Avoid running, except in an emergency.
- Take advantage of sounds that may distract the enemy.
- Fall silently without making an outcry.

Walking

When walking at night—

- Place the heel down first. Balance the weight of the body on the rear foot until a secure spot is found.
- Lift the forward foot high to clear any stiff grass, brush, or other obstruction.
- Continue to balance body weight on the rear foot, lower the forward foot gently, toe first, to explore the ground for objects that might make noise. Step over fallen logs and branches, not on them.
- Lower the heel of the forward foot slowly; gradually transferring body weight to that foot.

Creeping

The low crawl and high crawl are not suitable at night when very near the enemy because an easily heard shuffling noise results. Creeping is the recommended method of movement:

- Creep at night on the hands and knees.
- Use your hands to feel for twigs, leaves or other substances that might make a noise. Clear a spot to place your knee. Keeping your hand at that spot,

bring your knee forward until it meets your hand. Then place your knee on the ground and repeat the action with the other hand and knee.

- Lay the rifle on the ground at your side and clear an area for it. Lift the rifle up and move it forward. Movement is slow and tedious, since it must be done silently.

Hitting the Deck at Night (Right-Handed Shooter)

To safely hit the deck at night from the standing position—

- Advance your left leg, place the butt of the rifle in your right armpit with the hand remaining on the pistol grip, and grasp it with the right hand at the balance.
- Quietly drop down on the right knee and left hand.

- Move the left leg carefully to the rear, and then move the right leg to the rear.

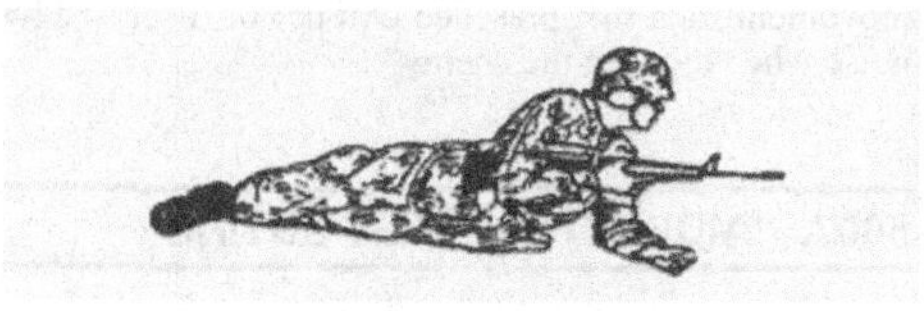

- Lie flat on the ground, or take up a firing position if necessary.

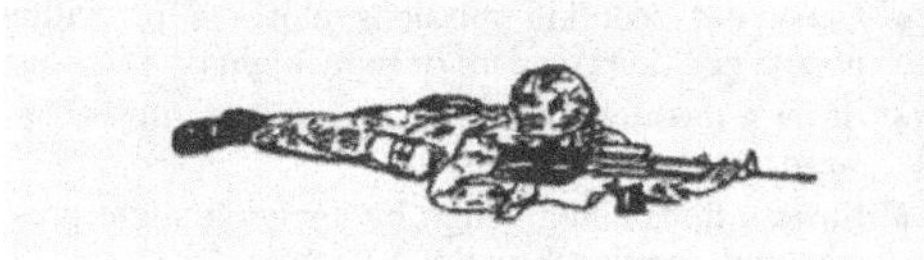

Wire Obstacles

A mission often requires a scout to pass through and work behind enemy positions. To accomplish this, the scout must be able to quietly pass through enemy wire obstacles and cross trenches. Cutting a gap in wire is time-consuming. If possible, walk over the low bands of enemy wire and crawl under the high bands (see fig. 5-5). Avoid movement along wire barriers, as enemy

Figure 5-5. Crossing Wire Silently at Night.

covering fires are generally planned parallel to them to take advantage of canalization and enfilade fire.

To step over low wire at night, crouch low to view the strands against the sky. Grasp the top strand with one hand; with the other hand, reach forward and feel for a clear spot for foot placement without stepping on other strands or any object apt to make a noise. Raise the body up, still grasping the top strand of wire. To avoid catching the foot in another strand, lift the foot up and over, passing it close to the hand grasping the wire.

If a high wire obstacle is encountered at night and wire cutters are not available, pass under the wire with your back on the ground. Grasp the lowest strands in your hands and hold them clear of the body while you slide under them.

When cutting wire and working solo, cut a wire near a post (see fig. 5-6), then dispose of all but one loose end. Grasp the wire close to a post and cut between your hand and the post, muffling the sound and keeping the loose wire in your grasp. When cutting wire in tandem, one firmly holds the wire with the hands positioned close to the cutters, in order to muffle the sound and prevent the loose ends from flying back, while the other one cuts. In both instances, the loose ends of the wire are bent back to form a passage.

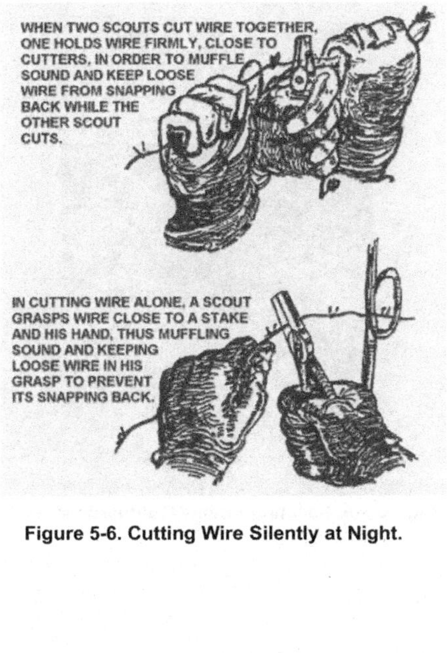

Figure 5-6. Cutting Wire Silently at Night.

Wrap a sandbag around the wire cutters and wire to deaden the sound.

Do not cut a complete gap in the wire; cut only the bottom wire(s). Leave the top wire(s) intact to lessen the chance of discovery by the enemy.

Crossing Trenches

Before approaching a trench, wait outside the trench for awhile and listen. Do not enter or cross a trench near its junction with a communication trench. Crawl silently up to the edge of the trench and look into it. Remove all loose dirt and rocks from the edge. If it is a narrow trench, spring up and jump across, sinking quietly to the ground on the other side and remaining there a moment to listen before proceeding. If the trench is wide, climb silently and slowly down into it and out the other side, using the revetment for support (see fig. 5-7). Do not enter enemy trenches unless it is absolutely necessary in order to accomplish the mission. Ordinarily, work is better accomplished from outside the trench. Sentries usually pay more attention to sounds in front of them; therefore, if it is necessary to enter a trench, cross it first at the place where enemy observation is restricted, then approach from the rear.

5009. LOCATING AND PLOTTING THE ENEMY AT NIGHT

For night work, a scout must understand the use of a lensatic compass. (Refer to para. 2002.) Using the lensatic compass, the scout can guide the platoon into position, locate adjoining elements of the command, keep direction when on patrol, determine the location of gaps in the enemy wire and the position of enemy out guards.

Locating Gaps in Enemy Wire

When searching for gaps in enemy wire, at least two lensatic compasses are needed: one to register the gap in the wire and the other for navigation. When a gap in enemy wire is located, lie outside the gap, keeping a distance of 10 meters from the barbed wire. Sight with the lensatic compass on a prominent point on the skyline in line with the gap. In selecting the prominent point in the skyline, pick one that appears on the map (i.e., hill mass, house, road junction). If the only prominent point available is one not identifiable on a map (i.e., a tree, destroyed vehicle, enemy position),

Figure 5-7. Crossing Trenches Silently at Night.

use it. The exact location of this point can be fixed the next day by visual reconnaissance of the area from an observation point. When the needle rests—

- Clamp it in place by lowering the eyepiece to the closed position.
- Rotate the movable bezel ring until the luminous line is directly over the north end of the needle. The azimuth of the gap from the prominent point is now registered.

The compass is carried back without further adjustment. The azimuth setting can be recorded later on a map (see fig. 5-8).

Locating Enemy Out Guards

At night, locate enemy guards by their sounds and failure to maintain light discipline. As sounds of the enemy are heard, and/or observations of the enemy made, shoot an azimuth with one compass. When the needle comes to rest, clamp the compass by lowering the eyepiece to the closed position.

Figure 5-8. Locating a Gap in Enemy Wire at Night.

Note the time and nature of each sound, the estimated distance, and which compass was used to fix the location. Plot this data on a map then wait until the debrief to turn in the notes and compasses. See figure 5-9.

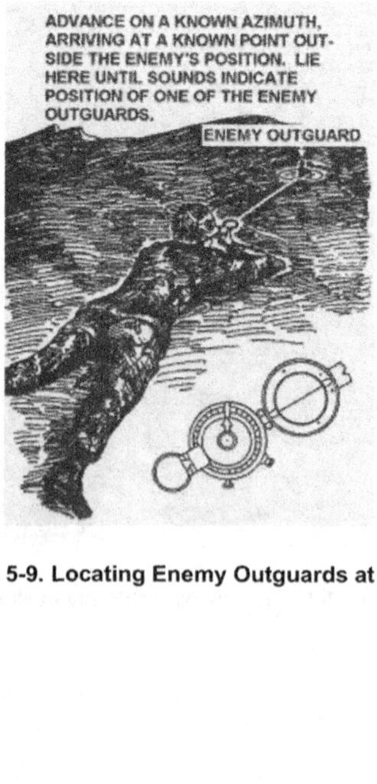

Figure 5-9. Locating Enemy Outguards at Night.

50010. ROUTES OF MOVEMENT

Prior to beginning a night missions, a scout studies the ground in detail from an observation point, air photos, and a map during daylight. The route of advance should be below the skyline. Avoid becoming a silhouette (see fig. 5-10).

Unless the moon is bright, avoid, if possible, passing through woods, ditches, ravines, and brush, because noises of movement may lead to discovery. If the enemy is known to have night observation device capability, avoiding these kinds of terrain may not be possible. To avoid enemy ambushes, return by a different route from the advance and change routes on successive nights.

Figure 5-10. Difference Between Correct Routes by Day or Night.

Chapter 6. Observing and Reporting

When sent out on a mission, a scout's duties are to observe and report. the first five chapters covered scout protection measures and movement in enemy territory. This chapter discusses performing the assigned mission and reporting the mission after its completion.

6001. OBSERVATION POSTS

Positions

The following guidelines apply to observation posts. See figure 6-1.

- When selecting observation posts, scouts should chose the least prominent position.
- Scouts may occupy one or more observation posts.
- An observation post should not be manned for more than 24 hours.
- A selected observation post should be observed for 10 to 15 minutes to ensure it is not occupied.
- Scouts move to the chosen observation post by a concealed route.
- If the post is located on a hill, crawl to a position where the skyline is broken.
- If a tree is used, the position should have a background so as not to be silhouetted against the sky while climbing or observing.
- When leaving the observation post, a different route from that of the approach should be used.
- If a radio is used, its antenna should be located to provide clear communication to the controlling commander but masked from enemy observation and direction-finding equipment. Upon departure, scouts should remove the antennae from the observation post so as not to give away the position.

Observing

Using all senses available, be particularly alert for movement, objects, sounds, and smells inappropriate to the surroundings. While observing, avoid all unnecessary movement. If observing from a building, keep back from doors and windows.

In daylight, look first at the ground nearest you. Begin observing close to your post and search a narrow strip 50 meters or less deep, going from right to left parallel to your front; then search from left to right a second and similar strip farther away but overlapping the first. Continue to observe until the entire field of view has been searched (see fig. 6-2 on page 6-2).

At night, use a night observation device. If one is not available, search the horizon with short, jerky movements, and short pauses. Look a little to one side of an object and then to the other. Lower the head close to the ground to view the object more clearly. Use low-powered field glasses to increase sight range.

Figure 6-1. Method of Approaching an Observation Post.

Figure 6-2. Method of Searching Ground.

6002. REPORTING

It is imperative that the scout accurately and completely report who, where, when, and what was observed to the proper authority upon mission completion. In most cases, a scout will not be equipped with a radio.

Verbal Reports

Verbal reports should be made when writing is impractical, when the information is not complicated or when the enemy is likely to intercept a messenger carrying a written message.

Written Messages

Written messages, preferred to verbal reports, are recorded in message book blanks issued for that purpose (see fig. 6-3) and delivered to a higher authority as soon as possible. The NATO spot report (SPOTREP) should be reviewed as the written message is a NATO format.

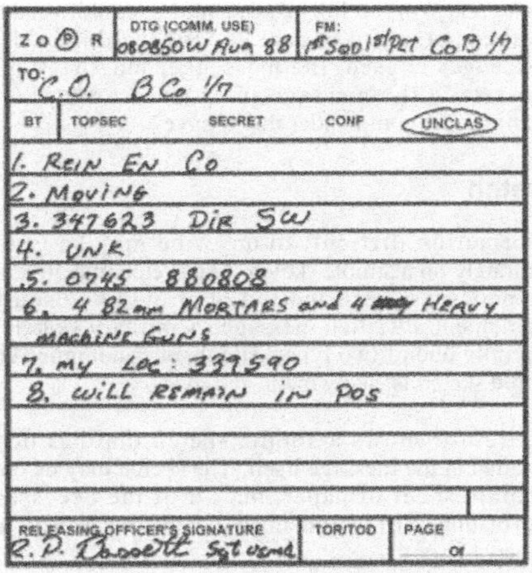

Z O (P) R	DTG (COMM. USE) 080850W AUG 88	FM: 1stSGD 1stPLT Co B 1/7
TO: C.O. B Co 1/7		
BT	TOPSEC SECRET CONF	UNCLAS

1. REIN EN CO
2. MOVING
3. 347623 DIR SW
4. UNK
5. 0745 880808
6. 4 82mm MORTARS and 4 HEAVY MACHINE GUNS
7. MY LOC: 339590
8. WILL REMAIN IN POS

RELEASING OFFICER'S SIGNATURE	TOR/TOD	PAGE OF
R.P. Bassett Sgt USMC		

Figure 6-3. Sample Field Message.

The message body is brief, accurate, and clear; facts and opinions are distinguished. If secondhand information is reported, its source is included. Reports include all information of value, first about the enemy, and then about the originator. Information about the enemy should cover—

- Size and/or strength.
- Actions or activity.
- Location and direction of movement.
- Unit identification. (The designation of the enemy unit may be derived from unit markings, uniforms worn or through prisoner interrogation.)
- Time of observation.
- Equipment and weapons.

Messages are printed in block letters. Individual items of information are numbered and separated into paragraphs. If doubt exists as to message receipt by the commander, a summary of its contents is included in the next message. Information about the originator or writer should cover—

- Location at the time of enemy observation (reference to an important terrain feature, by map coordinates, by the back azimuth from each of two definitely located points, or the back azimuth and distance from one known point).
- Intentions. (Remain in position? Continue on the mission? Take other action?)

The message is carefully reread and if possible read by another person to ensure understandability. If a messenger is used, the messenger must read and understand the message in order to answer any questions the commander might have.

Sketch

Information difficult to describe may be given accurately on a simple sketch. The sketch may give all the necessary information or it may be used to supplement a written message. A military sketch is generally one of two types: simple or panoramic. The simple sketch is easily made and read.

Figure 6-4 shows a simple sketch that has been included in the message itself. The sketch may be on a separate sheet of paper, but all of the necessary information must be contained in either the sketch, the message or both.

The panoramic sketch is a picture of the terrain's elevation in perspective, as seen from one point of observation. Although a panoramic sketch is not difficult to create, skill and training are necessary to enhance usefulness. Figure 6-5 on page 6-4 illustrates how to make a panoramic sketch.

Overlay

The same information sent back to higher headquarters on the sketch may be sent on an overlay, if the sender and the person to whom the message is to be sent have copies of the same map. Figure 6-6 on page 6-6 illustrates a simple overlay. The overlay is drawn on transparent paper as follows:

- Orient the map and place it on a hard, flat surface.
- Place the transparent paper over the part of the map of the object or information to be transmitted and hold the paper in this position.
- Orient the overlay to the map by tracing in the intersecting grid lines at two opposite corners of the overlay. Write the correct number designation on the overlay. The cross made by the intersection is called a *tick mark* and enables the receiver to locate the exact area on the map covered by the overlay.
- Sketch the object seen or the information to be transmitted on the tracing paper (the sheet on top of the map) in the exact location it would appear on the map (the sheet underneath the tracing paper).

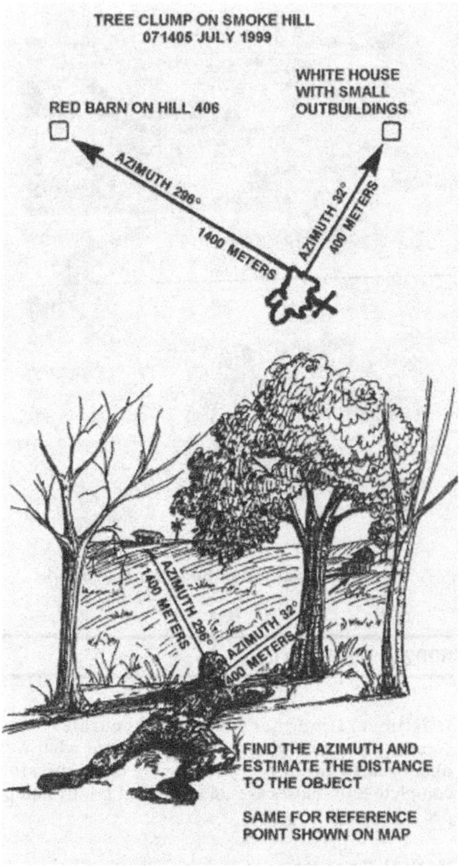

Figure 6-4. Making a Simple Sketch.

Explanatory notes are annotated in the overlay's margin, arrows point to the objectives.

- Indicate with an "X" and an appropriate explanation the position from which the observer saw the object or obtained information.
- Include title and scale of the map from which the overlay was made, date and hour the information was obtained, and signature of the observer on the overlay in the lower right-hand corner.

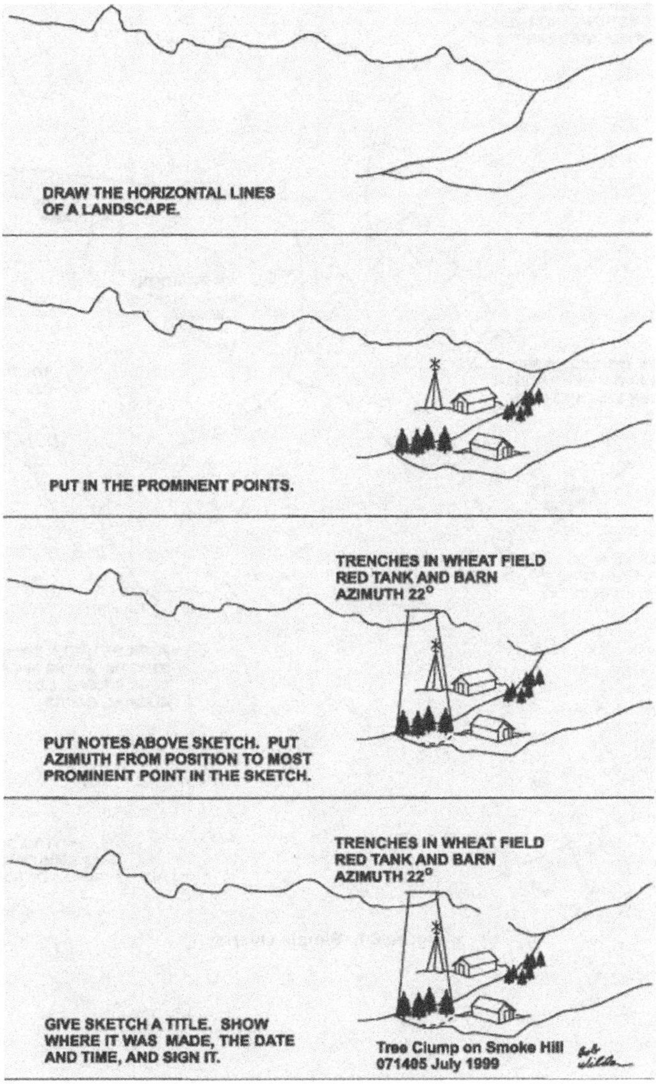

Figure 6-5. Panoramic Sketch.

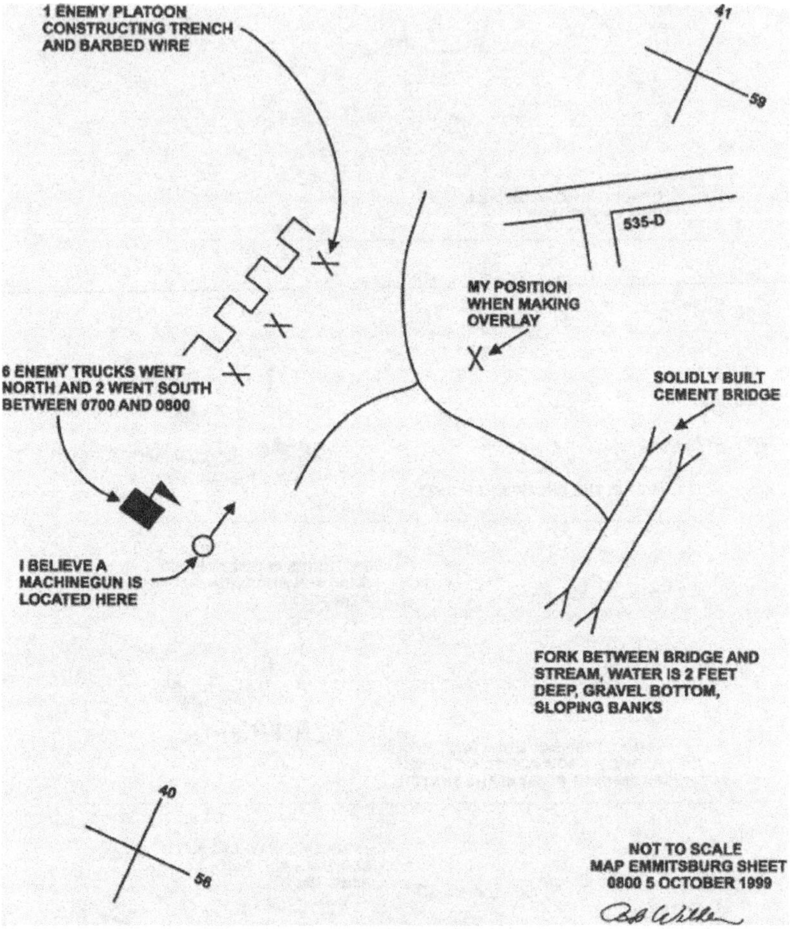

Figure 6-6. Simple Overlay.

CHAPTER 7. SCOUTING FIRE TEAMS

Upon completion of individual training, a Marine adapts this training to the functioning of the unit. The basic tactical unit for scouting is the fire team. Within the fire team, scouts normally work in pairs to furnish security and gather information. Information gathered by the individual scouts is passed to the fire team leader.

7001. POSITIONING

When a rifle platoon in the approach march is not preceded by adjacent units (left, right, front, and rear), it employs its own scouting elements. The scouting element for a platoon is usually one fire team; however, an entire squad may be used.

A fire team used as a scouting element is called a scouting fire team and is controlled by the platoon commander, assisted by the squad leader. A squad leader whose squad is providing the scouting fire teams normally marches near the platoon commander to assist in the control of the scouting fire teams.

A scouting fire team moves aggressively to cover the front of the advancing platoon and to locate the enemy's position(s). It generally moves in a wedge or skirmisher's formation. Normally, a scouting fire team is deployed on a frontage of 50 to 75 meters (10 to 17 meters between each individual scout). The entire squad may be employed to cover a wider frontage. The platoon commander coordinates the movement of the scouting fire team(s) so as to protect the main body of the platoon from enemy fire from points within 400 to 600 meters away, or in close terrain from points within the limits of enemy observation.

Scouting fire team(s) should have enough firepower to overcome resistance from small enemy advanced posts and patrols; the intent is to make enemy riflemen and machine gunners open fire and disclose their positions. Without scouts in advance, the platoon may move into areas where enemy fire may prevent further advance or maneuver and inflict heavy casualties (see fig. 7-1).

Scouting fire teams are covered by the platoon or, when the platoon is masked, the fire team leader ensures individual scouts are maneuvered and coordinated so that the fire team covers its own advance. The fire team leader constantly watches for signals from the platoon commander and remains in visual contact at all times.

The distance between the scouting fire team and the platoon is terrain-dependent. The scouting fire team should not be beyond visibility of the platoon. In open terrain, the platoon commander usually directs the scouting fire team to move by bounds along a succession of locations designated by the platoon commander as intermediate objectives.

Individual scouts should advance as stealthily as possible, while remaining consistent with their mission of reconnaissance to the front, taking advantage of cover without delaying the advance. An occasional glimpse of scouts constantly advancing over a wide front can make the enemy uneasy. It is this activity, and not the target the scouts offer, that may cause the enemy to open fire and disclose its location.

When fired upon, scouts must drop to cover and return fire only when necessary to complete the mission. If

Figure 7-1. Position of Scouts Preceding an Attacking Platoon.

necessary, one scout reports back to the fire team leader who informs the platoon commander.

7002. LOCATING ENEMY POSITIONS

The key terrain of defensive positions are those points that afford extended observation over the ground where the attack must advance. The enemy will place machine guns and infantry to defend critical points.

Members of a scouting fire team preceding an attacking platoon identify the probable enemy infantry and machine gun positions (see fig. 7-2). They use concealment and cover to conduct their advance in order to discover the exact location of enemy positions.

7003. ACTION WITH AN ATTACKING PLATOON

The scouting fire team reconnoiters to the front of the advancing platoon. As soon as the scouting fire team leader indicates the area is secure, the platoon advances and the scouting fire team moves forward. Squads within the platoon advance by bounds; at least one squad is positioned to support the other(s) by fire. Successive positions along the line of advance are selected and designated by the platoon commander as intermediate objectives, and reconnoitered by the scouting fire team before occupation. By conducting proper reconnaissance, surprise by the enemy or movement in the wrong direction may be prevented.

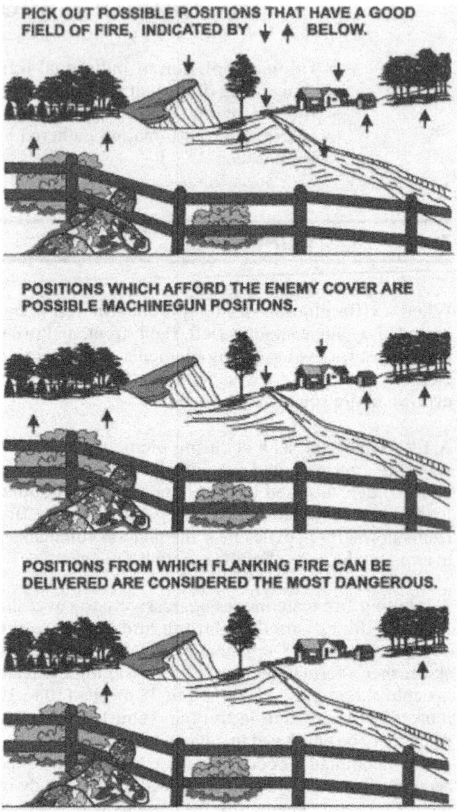

Figure 7-2. Assessing Probable Enemy Machine Gun Positions.

Movement

The distance between the scouting fire team and the front of the main body of the platoon is dependent on the mission, enemy, terrain and weather, troops and support available, and time available (METT-T). In close terrain, such as dense woods, the scouting fire team's movements closely resemble those used for night operations. In approaching houses, woods, and villages, one scout of each pair covers the other while the latter reconnoiters (see fig. 7-3).

A scouting fire team moves forward aggressively to cover the front of the advancing platoon, usually adopting either skirmisher's or wedge formation in order to be prepared to go into action immediately and to cover a wide frontage of 50 to 75 meters.

As a scouting fire team advances in open terrain, it is supported when possible by elements of the platoon; in close terrain, by mutual support within each fire team. Mutual support within the fire team is accomplished by the fire team leader and the automatic rifleman forming a team that supports by fire the advance of the rifleman and assistant automatic rifleman until they reach a location designated by the fire team leader. The rifleman and assistant automatic rifleman then support by fire the movement of the team leader and automatic rifleman. These successive points to which the sub elements of the fire team move are normally designated as fire team intermediate objectives by the fire team leader. The fire team leader sets as many fire team intermediate objectives as necessary to maintain mutual support within the team. This process is repeated until the team can be covered by other elements of the rifle platoon.

A scouting fire team takes advantage of available cover and concealment without delaying its advance. The orders of the platoon commander govern the distance at which it precedes the platoon. The terrain and the probable position of the enemy affect the scouting fire team's distance in front of the platoon. It may be as much as 400 to 600 meters in advance of the platoon. In open terrain, the platoon commander usually directs that the scouting fire team move by bounds to a succession of intermediate objectives. In

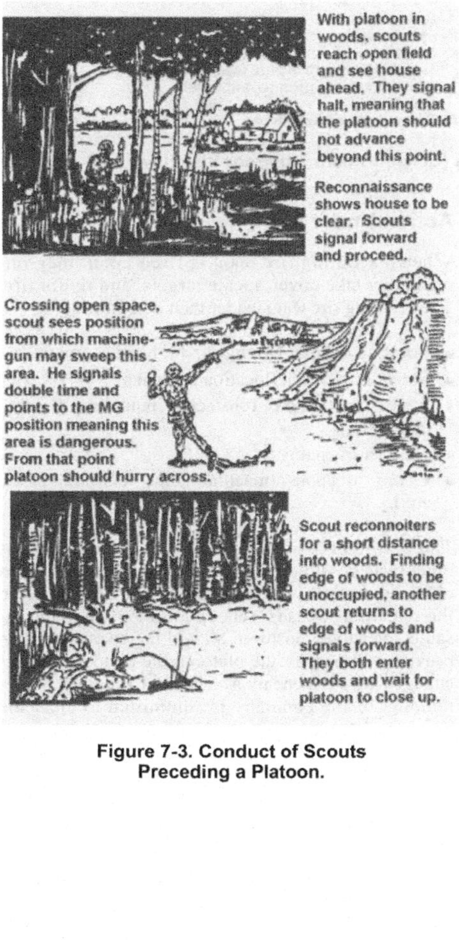

Figure 7-3. Conduct of Scouts Preceding a Platoon.

close terrain or conditions of limited visibility, the scouting fire team is normally ordered to precede the platoon at the limit of visibility, maintaining visual contact with the platoon commander.

Action in Woods

When a scouting fire team is directed to advance over open ground to the edge of a woodline, two members of the team, preferably the rifleman and assistant automatic rifleman, reconnoiter inside the woodline while the remainder of the fire team covers them. It is not recommended they separate until finished with their reconnaissance of the far side of the danger area. Both members staying together can cover the same area using a zigzag reconnaissance and they are better equipped to overcome any opposition.

In heavy underbrush and/or poor visibility, the rifleman and assistant automatic rifleman proceed into the woodline together for 50 to 60 meters. The two then separate, searching out either flank to the first high ground or limits of observation, probably 50 to 100 meters (see fig. 7-4 on page 7-4). After the initial search and out posting the limit of advance, the remaining scout signals the fire team forward.

In light underbrush and/or good visibility, the assistant automatic rifleman remains at the edge of the woodline while the rifleman searches the woodline. (see fig. 7-5 on page 7-5). The rifleman searches the woodline in a zigzag pattern, reports back to the assistant automatic rifleman, then moves to an outpost position at the limit of advance. In turn, the fire team leader signals the platoon commander that it is safe for the platoon to move forward.

The fire team leader then moves the remainder of the fire team into the woods, joining up with the forward scout manning the outpost. The scouting fire team occupies and holds a line 50 to 75 meters within the woods and observes toward the direction of movement until the platoon closes up. The scouting fire team leader awaits further word from the platoon commander before moving the team further into the woods.

When directed, the scouting fire team leader moves the team forward until they reach the far edge of the woods. The team is held at the edge of the woods and the fire team leader notifies the platoon commander of the situation. The platoon commander moves the platoon to a position where it can cover the scouting

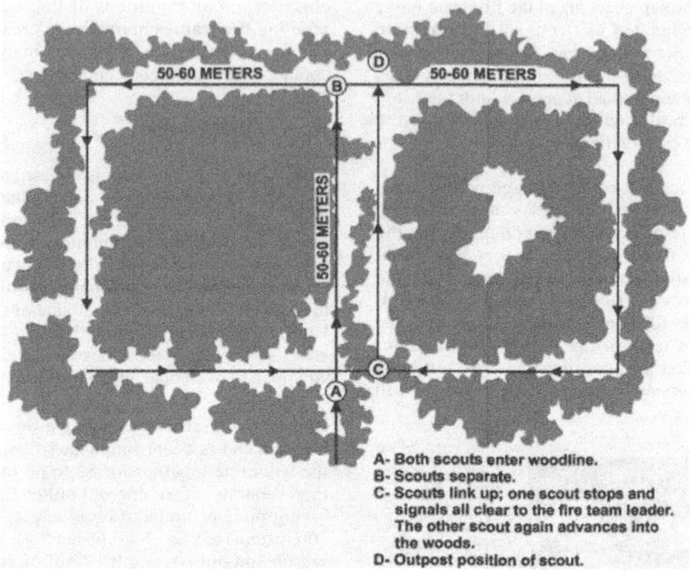

Figure 7-4. Searching Edge of a Woodline (Dense Underbrush/Poor Visibility).

fire team as it exits the woods and directs the team leader to move out and continue the scouting mission.

A scouting fire team passing through woods ahead of its unit maintains a distance allowing visual and oral communications. If an obstacle is encountered, reconnaissance to its front and flanks must be carried out. When advancing along a road or path, scouts precede the platoon to provide necessary protection and to prevent surprise fire on the platoon. When crossing a road or path, they reconnoiter well to the flanks before signaling "all clear" to the platoon.

The scouting fire team will not exit the woods until the arrival of the platoon commander, who will then be given an opportunity to alter the disposition or direction of march. The point where the platoon exits the woods is considered a danger area where the platoon is vulnerable to enemy fire. The scouting fire team is sent ahead to reconnoiter the danger area, as well as the next area to be occupied by the platoon. They signal back whether conditions require a halt, an advance or a quick rush across the open area. The scouting fire team leader must be continually on the lookout for signals from the rear.

Action Under Fire

When a scouting fire team is fired upon, they immediately take cover, locate targets, and return fire. The scouting fire team leader then determines—

- Location of enemy (range and reference points).
- Extent of position (location of flanks).
- Types of positions (obstacles, bunkders, fighting holes, etc.).
- Number of enemy.
- Enemy weapons (machine guns, mortars, tanks, etc.).

The platoon commander assesses the situation as quickly as possible based on the limited information obtained. Usually the platoon commander brings up the remaining squads, sets up a base of fire, and assaults the enemy position. Should the enemy position prove too strong for the platoon, the platoon remains engaged with the enemy as a base of fire until the remainder of the company is committed to clear the enemy resistance.

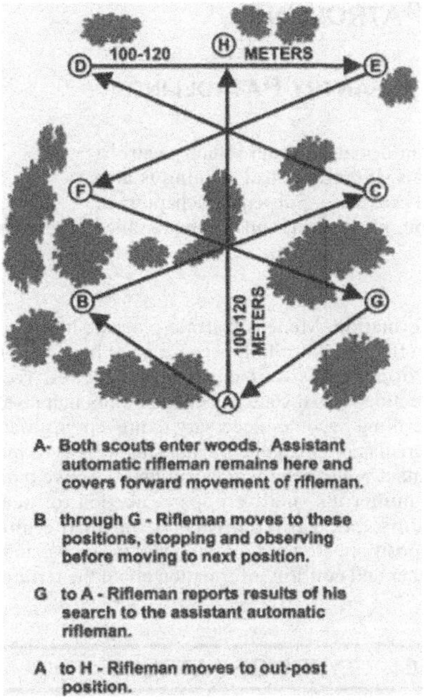

Figure 7-5. Searching Edge of a Woodline (Light Underbrush/Good Visibility).

7004. ACTION WITH AN ENVELOPING UNIT

When a platoon is given the mission to envelop an enemy position, a scouting fire team is employed for protection and reconnaissance in the same way as when the platoon is advancing in the approach march.

PART 2. INFANTRY PATROLLING

CHAPTER 8. FUNDAMENTALS OF INFANTRY PATROLLING

This chapter begins Part 2 and provides basic information about infantry patrols; specifically, their purpose, types, and missions. Infantry patrol training is also addressed and keys to successful patrolling are presented. Subsequent chapters of this part cover patrol organization, preparation, movement, and reconnaissance actions.

8001. DEFINITIONS

A patrol is a detachment of ground, sea or air forces sent out for the purpose of gathering information or carrying out a destructive, harassing, mopping-up or security mission (Joint Publication [JP] 1-02). The mission to conduct a patrol may be given to a fire team, squad, platoon or company.

8002. RELATION OF PATROLLING TO SCOUTING

Each patrol member must be knowledgeable in the principles of scouting and maintain membership of a larger team. To develop the teamwork skills required among the members of a patrol, additional training beyond the basic principles is necessary to become a well-trained scout. A patrol member must respond quickly to the decisions and orders of the patrol leader. There must be complete confidence among all members of the patrol and the confidence that they, as a team, will be successful in their mission.

8003. PURPOSE

A commander must have current information about the enemy and the terrain in order to employ the unit effectively. Patrols are an important means of gaining this information and are used to destroy enemy installations, capture enemy personnel, perform security missions or prevent the enemy from gaining information. Modern warfare places a high premium on effective patrolling because units have larger areas of operations and can be threatened from all directions. As distances between units increase, more patrolling becomes necessary to prevent infiltration by guerrillas or small enemy units, as well as to maintain contact with friendly adjacent units. Active patrolling by numerous small groups is needed to locate the enemy and gather information on the enemy's disposition, strength, morale, and weapons, as well as gather and confirm information about the terrain.

8004. TYPES OF PATROLS

Classification as to Mission

Reconnaissance Patrol

Reconnaissance patrols gather information about the enemy, terrain or resources. Relying on stealth rather than combat strength, they gather this information and fight only when necessary to complete the mission or to defend themselves. The distance covered by reconnaissance patrols varies based on the terrain and mission. The squad is ideally suited for reconnaissance patrol missions because of its relative small size and its experience of working together.

Combat Patrol

A combat patrol is a fighting patrol assigned missions that require engagement with the enemy in combat. Larger and more heavily armed than reconnaissance patrols, combat patrols have a mission to capture enemy documents, provide security, and capture or destroy enemy equipment and installations. Such action is ordinarily followed by a return to friendly positions. Regardless of the mission, the patrol reports any information concerning the enemy and terrain acquired during the accomplishment of the assigned mission. There are four types of combat patrols: raid, contact, ambush, and security (normally conducted by a Marine rifle platoon). A rifle platoon reinforced with crew-served weapons is normally considered the minimum size for contact, economy of force or ambush patrols. In some situations, such as the capture of a small enemy outpost, a rifle platoon could conduct a raid. However, a raid is a complex mission and, due to the organization of a raid force (command, reconnaissance, assault, support, security, and reserve elements), a rifle company is normally the smallest force assigned to a raid.

Classification as to Means of Movement

Foot Patrols

Movement by foot is the most common means; however, there are inherent disadvantages. Foot patrols travel slowly and carry limited quantity and types of equipment and supplies. Range and area coverage is relatively restricted. Foot patrols also have apparent advantages in that they have fewer restrictions as to terrain that can be covered; are more difficult for the enemy to detect; provide thorough coverage within limits of range; and are generally not inhibited by weather.

Motorized Patrols

Where terrain and road networks permit, a motorized patrol overcomes the inherent disadvantages of the foot patrol. Mechanized forces require patrolling units that can keep pace with them. However, motorized patrols are restricted to certain types of terrain, and tend to bypass areas that may be advantageous to and occupied by enemy infantry.

Waterborne Patrols

Waterborne patrols move over seas, lakes, rivers and streams, canals, and other inland waterways. The water is either used as a medium of entry to an objective area or is the actual patrol route. Waterborne patrols are limited by the location of water routes in the terrain and tend to bypass areas that may be advantageous to and occupied by the enemy.

Helicopterborne Patrols

Where terrain is extremely difficult or the enemy situation precludes the use of vehicle or motorized patrols, helicopterborne patrols are a method or means to conduct a patrol.

8005. TRAINING

Training is essential to successful patrolling. Premature and unordered actions by members of the patrol destroy coordination and control. Leaders are trained to issue their orders calmly to inspire confidence and discipline, and to avoid misunderstanding. Patrol members must work together and fight as a team. Training should develop the following skills:

- Expertise in handling individual and special weapons, and familiarity with enemy weapons that may be captured.
- Recognize camouflaged personnel, equipment, and defensive positions; ability to pick up fleeting targets fire the rifle from any firing position.
- Understand fire discipline and, after weapon firing, immediately change position (see fig. 8-1).
- Quick and accurate observation skills, and the ability to recall and transmit clearly and briefly, both orally and in writing.

Figure 8-1. Changing Position After Firing.

- Recognize and quickly respond to improvised signals, visually or by sound.
- Ability to swim with weapon and equipment.
- Use issued or improvised camouflage suits and garnish helmet in order to blend with the surroundings. Smudge face, hands, and any bright surfaces of weapons and equipment with some substance, such as mud or charcoal, to prevent the reflection of light.
- Silence self, equipment, and weapon.
- Use antimalarial and water purification tablets.
- Acclimation to temperature extremes.
- Develop a sense of direction and learn how to follow a course by compass, stars, sun, flow of streams, prominent terrain features, and by observing other natural phenomenon. Learn to determine the distance traveled from a known point and to keep a record of azimuths and the distance traveled on each azimuth (dead reckoning).
- Call for and adjust indirect fire assets.
- Familiarization with all communications assets and the use of field expedient antennae.

8006. KEYS TO SUCCESSFUL PATROLLING

Regardless of the category or means of conducting a patrol, the keys to successful patrolling are—

- Detailed planning. Every portion of the patrol must be planned, all possible contingencies considered.
- Productive, realistic rehearsals. Each phase of the patrol is rehearsed, beginning with actions in the objective area. Similar terrain and environmental conditions are used when conducting rehearsals.
- Thorough reconnaissance. Ideally, the patrol leader will physically conduct a reconnaissance of the route and objective. Photographs and/or maps will be used to supplement the reconnaissance.
- Positive control. The patrol leader must maintain positive control, this includes supervision during patrol preparations.
- All-around security. Security must be maintained at all times, particularly near the end of the patrol where there is a natural tendency to relax.

CHAPTER 9. PATROL ORGANIZATION

Organizing a patrol is a two-step process: the general organization of the entire patrol and the task organization of various patrol elements. Normally, the nature of patrolling does not permit long preparation periods and rehearsals to specifically build a unit for each mission. Accordingly, the patrol leader must combine unit integrity considerations with proven concepts of patrol organization.

9001. GENERAL ORGANIZATION

The patrol leader establishes a patrol headquarters and elements to accomplish the mission.

The headquarters is composed of the patrol leader and the personnel who provide support for the entire patrol, such as a forward observer, corpsman, and radio operator.

The major subdivisions of reconnaissance and combat patrols are elements. The existing infantry structure (squads and fire teams) is reinforced as required. For example, a reinforced platoon tasked to conduct a combat patrol that will raid an enemy outpost could be organized as follows:

- Platoon headquarters (command element):
  - Patrol leader (platoon leader).
  - Assistant patrol leader (platoon sergeant).
  - Navigator.
  - Radio operator (company tactical net).
  - Radio operator (patrol tactical net).
  - Corpsman.
- The first squad (security element) provides security en route to the objective area (the point, flank security, and rear security) and at the objective area (the flanks and the objective rally point).
- The second squad (support element) supports by fire for the attack, covering fire for the withdrawal, and supporting fires to cover the crossing of danger areas.
- The third squad (assault element) provides the assault force to attack and seize the objective; searchers to clear the objective; pacers, compass man, navigator, and the assistant patrol leader en route and back from the objective area.

Any attachments a patrol may have (i.e., demolition team, scout snipers, and machine gun squad) will be added to the element that supports its function. For example, the demolition personnel should go with the unit conducting the attack, and scout snipers and machine gunners should stay with the support squad.

9002. TASK ORGANIZATION

The preceding paragraphs described the elements necessary for a patrol to accomplish its mission. These elements reflect the internal functions or tasks required for the patrol to succeed. Depending on the METT-T, there are various methods of grouping these elements together. Task-organization is the further subdivision of patrol elements into teams that are required to perform essential tasks. In creating teams, unit integrity of infantry units should be maintained.

The patrol is organized so each individual, team, and element is assigned a specific task, but capable and prepared to perform other tasks. This may not be possible for certain specialist tasks requiring a trained technician.

CHAPTER 10. PATROL PREPARATION

For a patrol to succeed, all members must be well trained, briefed, and rehearsed. The patrol leader must have a complete understanding of the mission and a thorough understanding of the enemy and friendly situations. The patrol leader should make a complete reconnaissance of the terrain to be covered (either visual or map), and must issue an order to the patrol, supervise preparations, and conduct rehearsals.

10001. MISSION

The mission assigned to a patrol must be clear and oriented toward one objective with a specific task and purpose. More than one primary objective or indefinite missions invites confusion, casualties, and failure.

10002. FACTORS INFLUENCING PATROL SIZE

The circumstances under which patrolling by infantry units is conducted make it necessary for combat patrols to be able to fight offensively, security patrols to defend themselves, and reconnaissance patrols to move quickly and only fight if necessary to break contact or defend themselves. The size of a patrol depends on METT-T.

Generally, a patrol should contain the least number of members needed to accomplish the mission. Combat missions ordinarily require larger patrols than reconnaissance missions.

10003. COMMANDER'S DUTIES

Determining Patrol Requirements

The need for conducting patrols derives from the commander's stated mission (issued by higher level commander) and other specified and implied tasks (secondary or supporting efforts necessary to accomplish the stated mission). This requires analyzing the unit's mission and determining the necessary reconnaissance and/or combat tasks that must be performed. By considering the mission, along with time available, the commander develops the overall concept of operations to include the patrol plan and the specific tasks for each patrol.

Assigning Units

When assigning patrol missions, maintaining the integrity of the existing unit while considering the skills and experience of the unit and its leader are critical factors to the infantry commander. To provide operational depth and equitable apportionment of hazardous assignments, the commander ensures that each of the subordinate leaders and units develops the skills and experience necessary to conduct successful patrols. Assignment of patrol units must consider the commander's concept of operations as a whole and the plans for subsequent employment of assigned forces after completion of the assigned patrolling mission.

Providing Adequate Time

The commander must allow the patrol sufficient preparation time by completing the mission analysis, estimate, and preparation of orders. The commander should use warning orders to alert subordinates to possible requirements and afford them the opportunity for concurrent planning.

Providing the Patrol Leader Information

The following information should be provided:

- A simple, straightforward explanation of the mission, particularly for night patrols.
- General routes (defined by checkpoints) or exact routes (defined by avenues of approach or other terrain features) to follow.
- Enemy composition, disposition, and strength.
- Location and activities of friendly troops.
- Outposts or other security elements through which the patrol is to pass.
- Terrain conditions.
- Missions and routes of other patrols.

- Time patrol is to depart and return.
- Method of reporting information while on patrol (radio, messenger), place where messengers are to be sent, and place where the patrol leader is to report upon completion of the patrol.
- The challenge and password to be used for exiting and reentering friendly lines.
- Special instructions such as locations to be avoided and essential elements of information higher headquarters is seeking.
- Report barrier/obstacle plan of friendly units if known; support available from friendly unit (e.g., medical evacuation [MEDEVAC], react force).
- Fire support available.

Providing Special Skills and Equipment

There will be situations when the unit assigned to conduct a patrol does not have the necessary technical skills or equipment organic to it to successfully accomplish the patrol mission. In these cases, additions (attachments) are made to the unit, such as—

- Machine gun and/or shoulder-launched, multipurpose assault weapon (SMAW) team and/or squad.
- Forward observers (mortar/artillery).
- Radio operator(s).
- Combat engineers.
- Tracked vehicle crewman to assess the ability to traverse the terrain.
- Corpsman.
- Other personnel (snipers, translators) as required.

Providing Miscellaneous Support

The commander must ensure that the patrol leader is provided with the food, water, ammunition, radios and batteries, maps, special clothing, and any other items required by the unit (includin attachments) needed for the mission. Post-patrol support such as debriefings by intelligence personnel must also be planned.

Reviewing the Patrol Leader's Plan and Preparations

Once the patrol leader receives the mission, conducts visual and/or map reconnaissance, and develops the plan, the commander confirms the patrol leader's understanding of the mission and plan for accomplishing it. This discussion between the patrol leader and commander ensures that the patrol leader understands the commander's intent and is properly prepared to provide the patrol leader an opportunity to ask for clarification or additional support, if required.

Debriefing the Patrol

Upon return of the patrol, the commander receives the patrol report at a debriefing attended by the patrol leader and all patrol members. The debriefing should be conducted as soon as possible following the patrol's return, while information is still fresh in the minds of the patrol members. A patrol report, based on the information collected during the debrief, is generated and forwarded to the next higher com-mander.

10004. PATROL LEADER DUTIES

The patrol leader organizes and prepares the patrol by using the six troop-leading steps to make the best use of resources available. These steps are to—

1. Begin planning.
2. Arrange for reconnaissance and coordination.
3. Make reconnaissance.
4. Complete the plan.
5. Issue the order.
6. Supervise.

Begin Planning

The patrol leader begins by evaluating all factors affecting the mission. He looks for possible courses of action that lead to a decision, and then transforms this decision into an order. The first step includes making an initial assessment and decision on using available time, issuing a warning order and initial preparatory tasks, and initiating his estimate.

The initial planning effort assesses the time, assistance, and information available, and plans the proper use of each. Time allowances include reconnaissance; completion of the estimate and order; troop preparation; and such briefings, rehearsals, and inspections as required before beginning the patrol.

The patrol leader reviews the mission and the attachments and/or support available and decides what preparatory efforts must begin immediately. Proper use of subordinates to manage these initial tasks during this period reduces preparation time and frees the patrol leader for proper planning and reconnaissance. A warning order is issued using a modified five-paragraph order format (situation, mission, execution, administration and logistics, and command and signal [SMEAC]) as a checklist. (Refer to app. A for specific information that a patrol warning order should contain within each checklist item.

Once the initial preparations are set in motion, the patrol leader begins estimating by analyzing the mission; considering the friendly, enemy, and operating environments; considering each course of action available against what the enemy might do; comparing the courses of action in terms of mission accomplishment, capabilities, and probable casualties; and choosing one that becomes the basis for concept and order. While the patrol leader should use and organize notes, estimates must be done quickly and accurately, particularly for immediate situations. To organize thinking, the patrol leader uses METT-T, which consists of:

- Mission—the mission assigned to the patrol and how it relates to the mission of the commander who is sending the patrol.
- Enemy—what is known or suspected of enemy presence and capabilities, habits and characteristics, and fighting techniques.
- Terrain and weather—including ground, vegetation, drainage, weather, and visibility.
- Troops and support available—friendly situation and support available.
- Time available—the constraints and impact of time on preparation and mission accomplishment.

The estimate begins with mission analysis, which is the most important part of the entire planning process. Here, the patrol leader considers the specified tasks of the mission assigned and identifies other significant actions (specified and implied tasks) that must be undertaken to accomplish the stated mission. These tasks are arranged in sequence of accomplishment. The tasks and sequence create the framework for developing courses of action for the patrol concept of operations.

Arrange for Reconnaissance and Coordination

The patrol leader arranges a personal reconnaissance to observe as far forward as possible and also coordinates with the appropriate commanders for the patrol's "passage of lines" (see paragraph 11001 for complete definition) and supporting fires. The patrol leader also coordinates with other patrol leaders who may be operating in the same or adjacent areas and requests that the commander assigning him the patrol mission coordinate the patrol action with adjacent commanders, local security, and night defensive fires, as appropriate. The patrol leader may delegate any or all of these arrangements to the assistant patrol leader if the patrol leader requires the time for planning.

Make Reconnaissance and Complete the Estimate

The patrol leader uses personal reconnaissance to answer questions that arise from the map reconnaissance and METT-T evaluation. Specific points include passage points, lanes through obstacles, locations of friendly listening posts and observation posts, possible approach and return routes, enemy positions (if any), and intermediate observation points on the way to the objective.

In selecting approach and return routes, the patrol leader chooses routes that best use concealment and avoid opposition and obstacles. To lessen the chances of ambush by the enemy, the return trip is planned along a different route. In addition to personal reconnaissance and review of the map and aerial photographs, the advice of other patrol leaders who already are familiar with the terrain and the objective area should be considered.

After compiling information about the situation and possible time constraint, the patrol leader completes an estimate. The first step is developing courses of action, each of which will provide for movement to the objective area, mission accomplishment, and the return, based on the tasks and their sequencing identified in mission analysis. While the eventual concept of operations is presented in order of occurrence, the patrol leader must develop the courses of action by either backward or forward planning. In situations where the objective is well defined and there is sufficient information to plan the action for mission accomplishment (reconnaissance or combat), the patrol leader begins the scheme for accomplishing the mission at the objective and then, planning backwards, considers the options for getting there and back.

The following sections on movement to and return from the objective area, reconnaissance missions, and combat missions discuss methods and options available to the patrol leader in developing the courses of action. The principal variables between courses of action will be who, where, and how in the following:

- Patrol task organization.
- Routes to the objective area.
- Observation point(s) (reconnaissance patrol), ambush site, form of maneuver, type of ambush (combat patrol), fire support plan.
- Return routes.

The patrol leader then mentally considers the progress of each course of action (a map or simple sketch is a useful aid) against expected and unexpected enemy action. By comparing the options against each other and prospective enemy opposition, the patrol leader chooses the course of action that has the best chance of success. Includes in this mental preview process is the time to determine the patrol plan for unexpected contingencies (enemy attack/counterattack, casualty handling). These contingency actions, together with the selected course of action, become the patrol concept of operations.

Once the patrol leader determines the scheme of maneuver, the fire support required to accomplish the mission is addressed including the fire power organic to the patrol and what additional indirect fire support will have to be provided by other units. When planning for indirect fire support, the patrol leader considers the following questions:

- Will artillery, mortar, or close air support be required at the objective area (combat patrols)?
- What artillery and mortar targets exist along the routes to and from the objective area that can be employed by the patrol if it encounters the enemy during movement (reconnaissance patrols and combat patrols)?
- What additional fire support will be required to cover the patrol's movement from the objective area back to the friendly area once the enemy is aware of the patrol's actions at the objective area (combat patrols)?

The effect that casualties have upon the patrol depends upon many factors. Generally, more casualties can be expected in a combat patrol than in a reconnaissance patrol. A patrol may continue on to the objective carrying its casualties, send them back with a detail of Marines, abort the mission and return the entire patrol with the casualties, or call their parent unit for assistance.

Some factors that determine what action the patrol leader takes are: patrol's mission; unit's standing operating procedure for handling wounded; number of casualties and nature of their injuries; availability of aid, helicopters or other means of casualty evacuation. Helicopter evacuation should only be used for the most serious casualties. For infantry units conducting patrols in proximity to the enemy, helicopter evacuation of casualties may compromise the patrol's mission and force the patrol to return to friendly positions before the mission is completed.

The patrol leader determines the requirement for nuclear, biological, and chemical (NBC) defense equipment. Gas masks should always be carried due to the availability of riot control agents (RCAs) to the enemy. If chemical or biological agents have been employed in the area that the patrol must pass through, protective garments will have to be worn by patrol members for part of or the entire patrol. Wearing extra clothing and carrying extra equipment affects the speed of the patrol's movement. A contingency plan for post patrol decontamination must be developed.

Complete the Plan

At this point, the patrol leader has completed the basic thinking necessary for accomplishing the assigned mission. The patrol leader prepares the patrol order to spell out the details, assign tasks to subordinates, and explain the entire endeavor for ease of understanding by the other members of the patrol.

Prepare the Order

The patrol leader's order contains more detailed information than discussed in the warning order. Orders follow the prescribed five-paragraph order format but contain greater detail. The patrol order is a modified 5-paragraph order; the major modifications are to paragraphs 3a and 3c. The format for the order is contained in appendix B.

Issue the Order

The patrol leader asks for a status report on the initial preparatory tasks assigned to subordinate leaders and specialists when the warning order was issued. When the patrol leader has completed planning and initial preparations have progressed to the point where the patrol order may be issued, the members of the patrol are assembled. Roll call is taken to ensure all patrol members are present, then the prepared order is issued.

This will be the only opportunity for the patrol leader to issue detailed instructions. The mission, in particular, must be unmistakably clear so that once the patrol is committed, all subordinate leaders can act with unity of purpose.

Whenever possible, the patrol leader should have a Marine, such as the navigator, build a terrain model using dirt, sand, twigs, etc., explaining the concept of operations for movement to the objective area, actions at the objective area, and the return. Terrain models provide patrol members with a clear and simple layout of the area of operations and key terrain.

Supervise

Inspections and rehearsals, vital to proper preparation, are conducted even though the patrol leader and patrol members are well experienced in patrolling. Inspections determine the patrol's state of physical and mental readiness.

The patrol leader inspects before rehearsals to ensure completeness and correctness of uniform and equipment. The following areas are checked:

- Camouflage.
- Identification tags, Geneva Convention cards.
- Prescribed equipment, weapons, and ammunition are available and serviceable.
- Tape and other items are used to "silence" equipment (prevent noise produced during movement).
- Items that could provide information to the enemy (e.g., letters and papers) remain behind.
- Unnecessary equipment and excess weight remain behind.

The patrol leader questions each patrol member to ensure the following is known:

- The mission, planned routes (primary and alternate), and the fire support plan of the patrol.
- The individual's role: what to do and when to do it.
- What others are to do and how their actions impact.
- Challenges and passwords, codes, reporting times, radio call signs, frequencies, and any other pertinent details.

There is usually a period of time between final rehearsal and departure. The patrol leader reinspects just before departure to ensure all equipment is still in working order and the unit is ready to embark on the mission.

Rehearsals ensure the operational proficiency of the patrol. Plans are checked and needed changes are made. The patrol leader verifies the suitability of equipment. It is through rehearsals that patrol members become thoroughly familiar with the actions to take during the patrol.

If the patrol will operate at night, both day and night rehearsals are conducted. Terrain similar to that over which the patrol will operate is used. All actions are rehearsed. If time is limited, the most critical phases are rehearsed. Action at the objective is the most critical phase of the patrol and is always rehearsed.

An effective method is to talk the patrol through each phase, describing the actions and having each member perform individual duties. When satisfied, the patrol leader walks the patrol through all phases of the patrol using only the signals and commands to be used during the actual conduct of the patrol. Rehearsals continue until the patrol is thoroughly familiar with the plan. The rehearsal is also used to test the soundness of the patrol order and patrol organization.

After the rehearsal, the patrol leader makes final adjustments to the plan and patrol organization based on what was learned during the rehearsal and from other sources, such as the S-2 and adjacent patrols. When this is completed, the patrol leader issues final instructions to subordinate leaders noting any changes made in the patrol organization or plan. While the subordinate leaders are briefing the remainder of the patrol members, the patrol leader reports to the commander stating that the patrol is ready to begin the mission. The patrol leader also coordinates the location and time that the patrol can test fire all weapons prior to departure.

Chapter 11. Movement to and Return from the Objective Area

This chapter provides guidance to patrol leaders for movement to and return from the objective area. Action in the objective area depends on whether the patrol is assigned a reconnaissance mission, combat mission or security mission. Chapters 12 and 13 provide detailed guidance on these types of missions.

11001. PASSAGE OF LINES

A passage of lines is an operation in which a force moves forward or rearward through another force's combat positions with the intention of moving into or out of contact with the enemy. (JP 1-02)

During the initial preparation for the patrol, the patrol leader selects a patrol assembly area and reconnoiters the area of passage designated by the commander. In coordination with the unit commander responsible for the area of passage, the patrol leader identifies gaps or lanes in minefields and wire obstacles and locates local security elements through which the patrol will pass. The patrol leader also checks the route from the patrol assembly area to the passage point or contact point where the patrol will depart friendly lines. If possible, both the route to the passage point and the route through the frontlines should be concealed from the enemy's view.

The patrol leader also reconnoiters the area for return passage of lines and coordinates with the unit commander responsible for the area of passage for passage points and lanes as necessary. The patrol leader observes these points from the direction that the patrol will use upon return to friendly lines, if possible, to aid in recognition upon return. The patrol leader provides the forward unit with information about the size of the patrol, general route, and expected time of return. The manner of challenge and recognition of the returning patrol should be coordinated in detail.

Upon return to friendly forward local security squads and/or frontlines, the patrol leader leaves the patrol in a covered position and moves forward with a radio operator and at least one Marine for security to make contact with the friendly unit in the manner agreed. After contact is made and recognized, the patrol leader rejoins the patrol and takes them to the passage point, personally checking in each member.

11002. ORGANIZATION FOR MOVEMENT

The patrol's task organization establishes the elements and teams needed to accomplish the mission in the objective area and to and from the objective. The patrol leader determines the formation(s) in which the patrol moves to the objective area.

Formations

The proper use of patrol formations is critical to the patrol's success. The squad and fire team formations described in FMFM 6-5 (proposed MCWP 3-11.2), *Marine Rifle Squad*, also apply to infantry patrolling formations. Because the movement of the patrol must be concealed from the enemy, the patrol normally moves through terrain that provides concealment. Control of the patrol in this type of terrain is difficult; thus the column formation, which is easily controlled, is normally used. However, as various types of terrain are encountered, the patrol leader uses the same considerations in determining the appropriate formation used in other infantry operations.

The standard squad and fire team formations are adaptable to a patrol. The patrol may change formations en route to match the situation and terrain. The patrol leader may have to sacrifice some control for better dispersion or give up some speed for greater stealth and more security. Other considerations include—

- Visibility, weather, terrain, and vegetation will influence dispersion and control of individuals and units. These factors may also affect the enemy; if visibility is good for the patrol, it is also good for the enemy. Two pieces of luminous tape worn on the back of the collar will aid in control and movement on dark nights. The collar is turned down when near the enemy. The tape can also be worn on the back of the cap, but should be covered or removed when near the enemy.

- Preserving the integrity of fire units (fire teams and, if attached, machine-gun teams) is of primary importance. If team members are detached from a machine-gun team, the unit integrity is lost and effectiveness is reduced. The patrol leader must also position fire units so as not to mask their fires.
- All-around defense of the patrol must not be sacrificed under any circumstances. The conventional squad and platoon formations provide adequate firepower in any direction required. When attachments are made to the patrol, the attachments are positioned within the formation to enhance the firepower of the patrol. If a fire unit, such as a machine-gun team or squad, is attached, it is incorporated into the all-around defense of the patrol by modifying the conventional formation (see fig. 11-1). It is permissible to employ machine-gun teams individually during movement. However, assault rocket squads and teams should be employed primarily as rifle fire units for movement. Firing the SMAW or AT-4 from within a patrol formation can present a serious back-blast danger to patrol members and should only be fired on the patrol leader's direction.
- Time allotted for mission accomplishment is also a major consideration. In selecting the formations, the patrol leader must consider the speed of the movement required to meet the time constraints (if any) imposed on the patrol. If required to meet a time schedule, a formation that permits rapid movement should be used. Speed, however, must never be permitted to force the patrol leader to make rash tactical decisions.

Exercise of Control

The patrol leader is positioned for best control over the patrol. The assistant patrol leader moves at or near the rear of the patrol. Other subordinate leaders move with their elements. All patrol members assist by staying alert and passing on signals and orders. A signal to halt may be given by any patrol member, but the signal to resume movement is given only by the patrol leader.

Arm-and-hand signals are the primary means of communication within a patrol and should be used exclusively when near the enemy. All members must know the standard infantry signals (refer to FMFM 6-5), as well as any special signals required, and be alert to receive and pass them to other members.

The patrol leader should speak just loudly enough to be heard. At night, or when close to the enemy, the patrol leader halts the patrol and has subordinate leaders come forward. They speak in a low voice and then pass the information to their subordinates by moving from member to member.

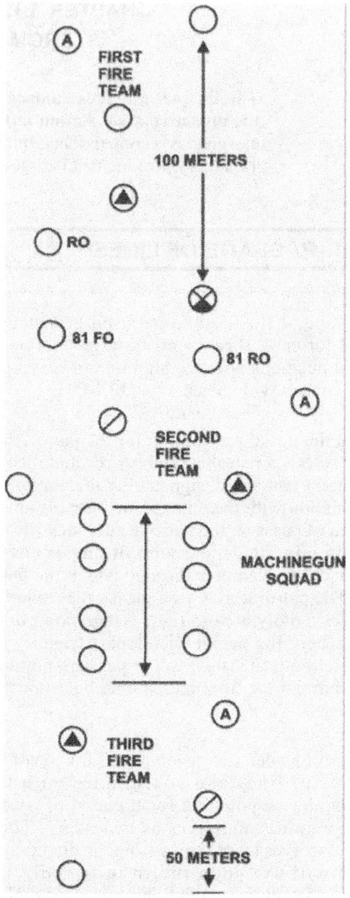

Figure 11-1. Squad Patrol with Attached Mortar Forward Observer Team and Machine Gun Squad.

Radios provide a means of positive control within a large patrol; however, radios should be used only when arm-and-hand signals or face-to-face contact between the patrol leader and subordinate leaders is impractical. When close to the enemy, words are spoken into the microphone with hands cupped over it in a low voice.

Other sound signals may be used if the patrol leader is sure they serve the purpose intended. Planned sound signals are rehearsed before starting on the patrol. Sound signals used must be simple, natural sounds that are few in number and easily understood. Bird and animal calls are seldom satisfactory.

Infrared equipment, such as the sniper scope and infrared filters for the flashlight, may be used as means of sending and receiving signals and maintaining control at night.

Luminous tape may be used to assist in control at night. Small strips on the back of the cap or collar of patrol members aid in keeping visual contact with the front member. However, the luminous tape must be covered when near the enemy.

An important aspect of control is accounting of personnel, especially after crossing danger areas, enemy contacts, halts, and exiting and re-entering friendly lines. The patrol leader may arrange for the last member to send up the count automatically after crossing danger areas, enemy contact, and halts. In large patrols or when moving in a formation other than a column, subordinate leaders check members and report the count to the patrol leader by the quickest method appropriate to the circumstances.

Navigation

One or more members are assigned as navigators for the patrol. Their function is to assist the patrol leader in maintaining direction by the use of the compass.

The patrol leader assigns at least two members as pacers to keep track of the distance from point to point. The average of their count is used to approximate the distance traveled. Pacers are separated so they do not influence each other's count. Pacers must know how to convert their own paces to meters.

The route is divided into increments, with each increment starting at a recognizable point on the ground. The pacers begin their counts from zero at the beginning of each leg. This makes the pace count easier to keep and checks for accuracy.

The pace count is sent forward when the patrol leader turns to the member behind and in a low voice says, "send up the pace" or uses the arm-and-hand signal of tapping his boot. This signal is passed to both pacers, who in turn send up the pace count in meters; for example, "two-hundred" or "one-seven-five."

Patrol members must understand that the counts of both pacers are sent forward. The patrol leader must know the counts of both men in order to check them.

Security

The patrol is dispersed consistent with control, visibility, cover, and concealment. Scouts are employed to the front, flanks, and rear of the patrol to provide security. For the patrol members traveling in the main body of the patrol, areas of responsibility are assigned to the front, flanks, and rear. Scouts are the eyes and ears of the patrol leader. They move when and where directed by the patrol leader and maintain contact with the patrol leader at all times, except when momentarily obscured by vegetation or other terrain features.

Front

Small patrols (squad size) may employ from one scout up to a fire team as the point, depending on the enemy situation, terrain, and route being followed by the patrol. Normally, squad-size patrols will use two scouts as the point. The point is responsible for investigating the route of advance immediately to the front of the patrol.

The point moves as far ahead of the patrol as visibility and terrain permit. When visibility is good, the point may precede the main body by as much as 100 meters. The point travels right and left ahead of the patrol, searching the area over which the patrol will pass.

The point maintains direction by knowledge of the general route to be followed and visual contact with the patrol leader. The patrol leader or the navigator ensures that the point is proceeding correctly.

The point, which stays far enough ahead of the patrol to provide security, is not a trail breaker for the patrol. If the point loses contact with the patrol, the point waits for the main body to catch up or moves rearward if contact is not quickly regained.

One of the navigators may be positioned with the point. One or more members works as the point while the other is the navigator.

Flank

Flank security for a patrol of squad size or less may be provided by using one or two members on either flank. If two scouts are assigned to a flank, one is positioned to observe the patrol leader and the other works farther out from the patrol. The scout who must observe the patrol leader remains within a maximum distance of 100 meters. The scout farther out remains in sight of the inside scout but normally does not move more than 20 or 25 meters away and remains prepared to relieve flank security regularly. Moving through dense woods or jungle may render the use of flank security impractical because of reduced visibility. In such cases, it moves with the patrol itself, but maintains observation to its assigned flanks.

Rear

A small patrol normally has only one rifleman assigned as rear security. An interval between the member assigned as rear security and the last Marine of the patrol is maintained at the limit of visibility, up to 50 meters. This member maintains rear security for the patrol by constantly observing to the rear.

Halting

Speed of movement is slower at night than in day patrols and reduces the danger of a Marine becoming separated from the patrol. The patrol occasionally halts to observe and listen for enemy activity; this is called a security halt. Upon signal, when reaching a danger area and periodically throughout movement en route, every member freezes in place, remains quiet, observes, and listens. It may be necessary to call a security halt just after departing friendly areas and just before entering friendly areas.

The patrol may halt briefly to send a message, eat, rest, check direction, or make a reconnaissance. The area selected should provide adequate concealment and cover, as well as favor the defense. All-round security is established and the patrol leader ensures all members move out when the patrol resumes movement. (For extended halts, see paragraph 11005.)

Infiltration

The disposition of enemy forces may sometimes prevent a patrol from entering the enemy occupied area as a unit; however, pairs of scouts or fire teams may slip through without being discovered. (Refer to FMFM 6-5 for infiltration techniques and procedures).

11003. CONTROL MEASURES FOR MOVEMENT

Checkpoints

A checkpoint is a predetermined point on the surface of the Earth used as a means of controlling movement, a registration target for fire adjustment or reference for location (JP 1-02). Checkpoints are a means of control between the parent unit and the patrol. These locations are decided upon and coordinated before the patrol leaves, so that both the patrol members and parent unit will know the patrol's location when it reports in. The parent unit can follow the progress of the patrol without transmitting coordinates over the radio that the enemy could monitor.

Rally Points

A rally point is an easily identifiable point on the ground where units can reassemble and reorganize if they become dispersed (MCRP 5-12C). It should provide cover and concealment, be defensible for at least a short time, and be easily recognized and known to all patrol members. All rally points are considered and identified as tentative rally points until they are occupied, found to be suitable, and designated as rally points by the patrol leader.

Initial Rally Point

The initial rally point is within friendly lines where the patrol can rally if it becomes separated before departing the friendly area or before reaching the first en route rally point. It may be the assembly area where the patrol waits while the patrol leader contacts the last friendly position through which the patrol will pass. The location of the initial rally point must be coordinated with the forward unit commander.

En route Rally Points

En route rally points are between the initial rally point and the objective rally point; and from the objective rally point back to the point where the patrol reenters friendly lines. They are determined as the patrol passes through a likely area that is suitable for a rally point.

Objective Rally Point

The objective rally point is located nearest the objective where the patrol makes final preparations prior to approaching the objective. It also serves as a location where the patrol reassembles after completing actions on the objective. The objective rally point must be suitable to accommodate those activities accomplished prior to actions on the objective. This position must provide the patrol concealment from enemy observation and, if possible, cover from enemy fires. It may be located short of, to a flank or beyond the objective. It should be out of sight, sound, and small-arms range of the objective area. The patrol leader's reconnaissance of the objective is made from this position; it is the release point from which patrol elements and teams move to the objective to accomplish the mission.

Rally Point Selection

The patrol leader selects likely locations for tentative rally points during reconnaissance or map study. A tentative rally point must be confirmed and announced after examination proves suitability. A tentative initial rally point and a tentative objective rally point are always selected and identified in the patrol order. If necessary, the patrol leader selects additional rally points en route as suitable locations are reached. When the patrol reaches a danger area that cannot be bypassed, such as an open meadow or stream, the patrol leader selects a rally point on both the near and far side. If good locations are not available, the patrol leader designates the rally points in relation to the danger area. The patrol leader will say, for example, "50 meters this side of the trail" or "50 meters beyond the stream."

Rally Point Use

The initial rally point and en route rally points are designated to enable the patrol to reassemble if it is unavoidably separated or dispersed. Identifying features are pointed out. The patrol leader ensures that the information is passed to all patrol members. When crossing a danger area, a near side rally point and a far side rendezvous point are designated. These rally points should only be used when all other methods of retaining control of the patrol have failed. The success of the patrol may be jeopardized if it is dispersed and cannot rally expeditiously.

If the patrol has left the friendly area and becomes dispersed, patrol members return to the last designated rally point (the initial or an en route rally point) unless the patrol leader gives other instructions.

As previously noted, the patrol leader selects two rally points at the near and far sides of danger areas that cannot be bypassed. If the patrol becomes separated or dispersed at a danger area, and there has been no enemy contact, the patrol should reassemble at the rally point on the far side of it. If, however, the patrol is separated or dispersed at a danger area as a result of enemy contact, members who have already crossed the danger area assemble and reorganize at the rally point on the far side; members who have not crossed assemble and reorganize at the rally point on the near side. In the absence of the patrol leader and unless directed otherwise, the senior member at the rally point on the near side takes charge, attempts to move the rallied patrol members to the rally point on the far side, and rejoins the remainder of the patrol.

Rally Point Actions

The patrol leader plans the actions to be taken at rally points and instructs the patrol accordingly in the patrol order. Planned actions at the initial rally point and en route rally points must provide for the continuation of the patrol as long as there is a reasonable chance to accomplish the mission. Plans for actions at rally points should provide for—

- Recognition signals for assembly at rally points.
- Minimum number of members and maximum amount of waiting time required before the senior member at the rally point moves the rallied patrol members onward toward the objective or returns to friendly lines.
- Instructions for patrol members who find themselves alone at a rally point.

11004. PRECAUTIONS AT DANGER AREAS

A danger area is any place where the patrol is vulnerable to enemy observation or fire (open areas, roads, trails, and obstacles such as barbed wire, minefields, rivers and streams, and lakes). Any known or suspected enemy position the patrol must pass is also a danger area. The patrol leader plans for crossing each danger area and includes these plans in the order.

The patrol reconnoiters the near side of a danger area first, then the patrol leader sends scouts to reconnoiter the far side. Once the scouts report that the far side is clear of the enemy, the remainder of the patrol crosses the danger area. As each individual or group crosses the danger area, they are covered by those remaining and by those who have successfully crossed. Enemy obstacles are avoided since they are usually covered by fire.

In crossing a river, the near bank is reconnoitered first; then the patrol is positioned to cover the far bank. Scouts are sent across to the far bank. After the far bank has been reconnoitered and the scouts report that it is clear of the enemy, the patrol crosses as rapidly as possible. This may be done individually or in pairs. If crossing the river requires swimming, the patrol uses improvised rafts to float equipment, weapons, and ammunition across. (Refer to MCRP 3-02C, *Water Survival Handbook.)*

A road or trail is crossed at or near a bend or where the road is narrow. Observation is restricted and, if the enemy is present, the patrol is exposed as short a time as possible. The near side is reconnoitered first, then scouts are sent across to reconnoiter the far side. This includes reconnoitering the tentative rally point on the far side. Once the scouts report "all clear," the remainder of the patrol crosses rapidly and quietly.

If the patrol must pass close to an enemy position, it takes advantage of battlefield noises to cover the sounds of movement. If supporting fires are available, the patrol leader can call for them to divert the enemy's attention as the patrol passes.

11005. HIDE

When a patrol is required to halt for an extended period in an area not protected by friendly troops, the patrol moves into a location which, by the nature of the surrounding terrain, provides passive security from enemy detection. Such an assembly area is termed a *hide*. To establishment a hide—

- Cease all movement during daylight hours to avoid detection.
- Hide the patrol for an extended period while the patrol leader conducts a detailed reconnaissance of the objective area.
- Rest and reorganize after extended movement.
- Reorganize after a patrol has infiltrated the enemy area in small groups.

The patrol leader's plan must include tentative hide locations when the patrol's mission dictates an extended halt within enemy areas. These tentative locations must be confirmed by actual ground reconnaissance prior to occupation by the patrol. The plan for a hide includes both passive and active security measures.

Passive security measures are—

- Avoid built-up areas.
- Select an area remote from all human habitation.
- Avoid known or suspected enemy positions.
- Avoid ridge lines, topographic crests, valleys, lakes, and streams.
- Avoid roads and trails.
- Avoid open woods and clearings.
- Select areas offering dense vegetation, preferably bushes and trees that spread out close to the ground.

Active security measures—

- Establish security covering all likely avenues of approach into the site.
- Establish communications (wire, radio, signal, runner) with posted security to provide early warning of enemy approach.
- Select an alternate area for occupation if the original hide is compromised or found unsuitable.
- Plan for withdrawal in the event of discovery.
- Establish an alert plan with a certain percent of the personnel awake at all times.
- Organize the elements of the patrol so necessary activities can take place with a minimum amount of movement.

The size of the area physically occupied by a patrol in a hide and the number of security posts required are governed by the terrain, quantity and quality of cover and concealment, and size of the patrol.

If the situation permits, a hide can also be used as the final preparation position and/or objective rally point.

11006. IMMEDIATE ACTIONS UPON ENEMY CONTACT

A patrol may make contact with the enemy at any time. Contact may be through observation, a meeting engagement or ambush. Contact may be visual, in which the patrol sights the enemy but is not itself detected. When this is the case, the patrol leader can decide whether to make or avoid physical contact, basing his decision on the patrol's assigned mission and capability to successfully engage the enemy unit.

When a patrol's assigned mission prohibits physical contact (except that necessary to accomplish the mission), its actions are defensive in nature. Physical contact, if unavoidable, is broken as quickly as possible and the patrol, if still capable, continues its mission.

When a patrol's assigned mission permits or requires it to seek or exploit opportunities for contact (as in the case of a combat patrol), its actions are offensive in nature, immediate, and positive.

In patrolling, contacts (visual or physical) are often unexpected at very close ranges, and short in duration. Effective enemy fire often provides leaders little or no time to fully evaluate situations and issue orders. In these situations, immediate action provides a means for swiftly initiating positive offensive or defensive action, as appropriate.

Two types of physical contact with the enemy are meeting engagement and ambush. Meeting engagement is a combat action that occurs when a moving force, incompletely deployed for battle, engages an enemy at an unexpected time and place. It is an accidental meeting where neither the enemy nor the patrol expect contact and are not specifically prepared to deal with it. An ambush is a surprise attack from a concealed position.

Immediate Actions

Immediate actions are designed to provide swift and positive small unit reaction to visual or physical contact with the enemy. They are simple courses of action in which all Marines are well trained. Minimal signals or commands are required and they are developed as needed for the combat situation. The signals can, in many cases, be initiated by any member of the unit. It is not feasible to attempt to design an immediate action drill to cover every possible situation. It is better to know the immediate action drill for each of a limited number of situations that may occur during a patrol. Arm-and-hand signals associated with immediate actions—such as FREEZE, ENEMY IN SIGHT, and HASTY AMBUSH RIGHT or LEFT—are contained in FMFM 6-5 (proposed MCWP 3-11.2).

Immediate Halt

When the patrol detects the enemy but is not itself detected, the situation requires the immediate, in-place halt of the patrol. The first member visually detecting the enemy gives the silent signal for FREEZE. Every member halts in place, weapon at the ready, and remains absolutely motionless and quiet until further signals or orders are given.

Air Observation and/or Attack

These actions are designed to reduce the danger of detection from aircraft and casualties from air attack.

When an enemy or unidentified aircraft that may detect the patrol is heard or observed, the appropriate immediate action drill is FREEZE. The first member hearing or sighting an aircraft that may be a threat signals FREEZE. Every member halts in place until the patrol leader identifies the aircraft and gives further signals or orders. Members of the patrol must not look up at the aircraft as sunlight can reflect off their faces even when camouflaged.

When an aircraft detects a patrol and makes a low level attack, the immediate action drill air attack is used. The first member sighting an attacking aircraft shouts, "AIRCRAFT," followed by the direction of the incoming attack: FRONT, LEFT, REAR or RIGHT. The patrol moves quickly into line formation, well spread out, at right angles to the aircraft's direction of travel. As each member comes on line, the member hits the ground, using available cover, then positions the body perpendicular to the aircraft's direction of travel, to present the shallowest target possible (see fig. 11-2 on page 11-8). Between attacks (if the aircraft returns or if more than one aircraft attacks), patrol members seek better cover. Attacking aircraft are fired upon only on command of the patrol leader.

Meeting Engagement

Hasty Ambush. This immediate action is used to avoid contact and to prepare to initiate an unplanned ambush on the enemy. It may often be a subsequent action after the command freeze. When the signal HASTY AMBUSH is given (by the point member, patrol leader or another authorized patrol member), the entire patrol moves quickly to the right or left of the line of movement, as indicated by the signal, and takes up the best available concealed firing positions (see fig. 11-3 on page 11-9). The patrol leader initiates the ambush by opening fire and shouting, "FIRE"; thus ensuring initiation of the ambush if the weapon misfires. If the patrol is detected before this, the first member aware of detection initiates the ambush by firing and shouting. The patrol leader may decide not to initiate the ambush in order to avoid contact unless the patrol is detected. When used as an offensive

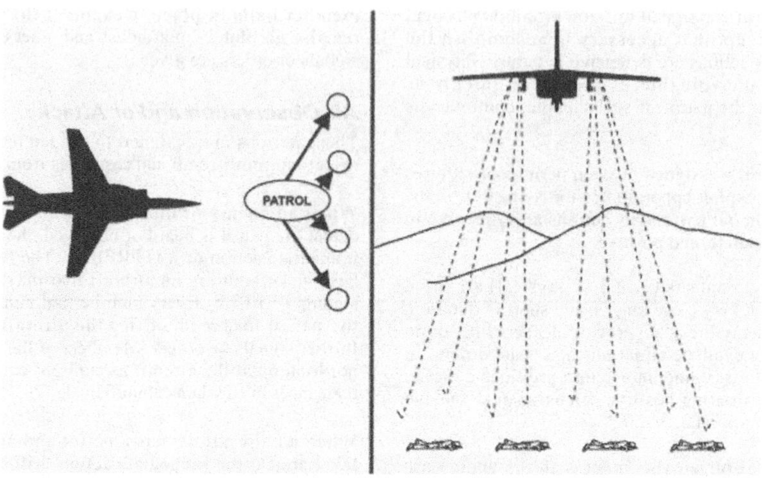

Figure 11-2. Immediate Action, Air Attack.

measure, the enemy is allowed to advance until he is in the most vulnerable position before the ambush is initiated. An alternate means for initiating the ambush is to designate an individual (for example, point or last member) to open fire when a certain portion of the enemy unit reaches or passes that member.

Immediate Assault. This immediate action drill is used defensively to make and quickly break undesired but unavoidable contact (including ambush) and offensively to decisively engage the enemy (including ambush). When used in a meeting engagement, members nearest the enemy open fire and shout, "CONTACT," followed by the direction of the incoming attack: FRONT, LEFT, REAR or RIGHT. The patrol moves swiftly into line formation and assaults (see fig. 11-4).

Defensive Measures

When used defensively, the assault is stopped if the enemy withdraws and contact is broken quickly. If the enemy stands fast, the assault is carried through the enemy positions and movement is continued until contact is broken.

Offensive Measures

When used offensively, the enemy is decisively engaged. Escapees are pursued and destroyed until orders to break contact are given by the patrol leader. If the patrol is fired upon from beyond 50 meters, the patrol must break contact as quickly as possible and continue the mission. If it engages the enemy any longer than necessary to break contact, it may put the mission in jeopardy.

Fire and Maneuver

Fire and maneuver is one means to break contact. One portion of the patrol returns the enemy fire while another portion moves by bounds away from the enemy. Each portion of the patrol covers the other by fire until contact is broken by all.

Clock System

The clock system is another means to break contact. Twelve o'clock is the direction of movement of the patrol. The patrol leader shouts a direction and a distance. For example: "TEN O'CLOCK-TWO HUNDRED," means the patrol should move in the direction of ten o'clock for 200 meters. Patrol members keep their same relative positions as they move so the original formation is not disrupted. Subordinate leaders must be alert to ensure that the members of their elements and teams receive the correct order and move as directed.

Counter Ambush

When a patrol is ambushed, the immediate action drill used depends on whether the ambush is a near ambush

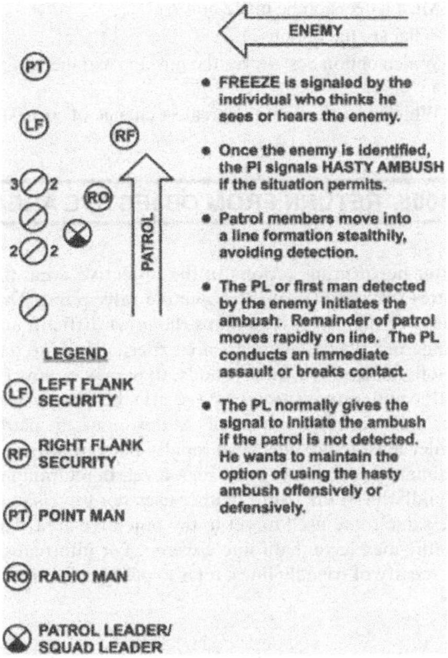

Figure 11-3. Immediate Action, Hasty Ambush.

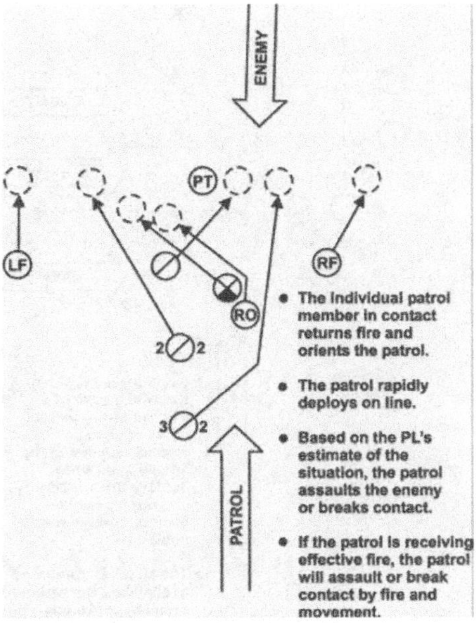

Figure 11-4. Immediate Action, Immediate Assault.

(the enemy is within 50 meters of the patrol) or a far ambush (the enemy is beyond 50 meters of the patrol). Fifty meters is considered the limit from which the ambushed patrol can effectively launch an assault against the enemy.

Near Ambush. In a near ambush, the killing zone is under very heavy, highly concentrated, close range fires. There is little time or space for members to maneu-ver or seek cover. The longer they remain in the killing zone, the greater the chance they will become casualties. Therefore, if members in the killing zone are attacked by a near ambush, they immediately assault without order or signal directly into the ambush position, occupy it, and continue the assault or break contact, as directed. This action moves them out of the killing zone, prevents other elements of the ambush from firing on them without firing on their own members, and provides positions from which other actions may be taken (see fig. 11-5 on page 11-10). Members not in the killing zone maneuver against the enemy as directed. The assault continues until all patrol members are outside of the killing zone.

Far Ambush. In a far ambush, the killing zone is also under very heavy, highly concentrated fires, but from a greater range. This greater range provides members in the killing zone maneuver space and some opportunity to seek cover at a lesser risk of becoming a casualty. If attacked by a far ambush, members in the killing zone, without order or signal, immediately return fire, take the best available positions, and continue firing until directed otherwise. Members not in the killing zone maneuver against the ambush force, as directed (see fig. 11-6 on page 11-10). The assault is continued against the enemy or until the order to break contact is given.

In each situation, the success of the counter ambush employed depends on the members being well trained in quickly recognizing the distance from which an ambush is initiated and well rehearsed in the proper reaction.

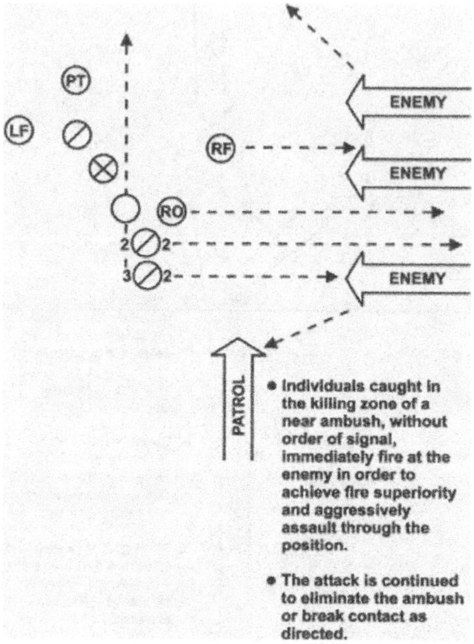

Figure 11-5. Near Ambush.

11007. PATROL LEADER'S ACTION IN A DEVELOPING SITUATION

While good patrolling depends on good planning, the patrol leader's plan must be flexible. Every combat situation develops differently than expected, and the patrol leader must be ready to quickly adapt to the situation as it develops. One of the most difficult tasks in battle is to recognize the correct moment for making a decision. Generally, it is more difficult to determine the moment for making a decision than it is to formulate the decision itself. When the situation demands, decisions must be made promptly without waiting for more information. In a developing situation, the patrol leader should use the following questions as a guide for battlefield decisionmaking:

- How has the situation changed?
- How does the change affect mission accomplishment and the immediate superior's mission?
- Must a decision be made now?
- What are the options?
- Which option best serves the mission and the unit as a whole?
- Which option offers the greatest chance of success?

11008. RETURN FROM OBJECTIVE AREA

After performing actions in the objective area, the patrol reassembles at the objective rally point. This phase of the patrol is perhaps the most difficult and dangerous. Patrol members are experiencing fatigue, emotional letdown, and wounds; they may be low on water and ammunition. Above all, the enemy was likely alerted if not in pursuit. At this point, the patrol leader must move the patrol rapidly but carefully and maintain patrol security at a high level. In returning to friendly lines, the patrol neither uses nor travels near the same route used to get to the objective area. The enemy may have that route covered. For information on reentry of friendly lines, refer to paragraph 11001.

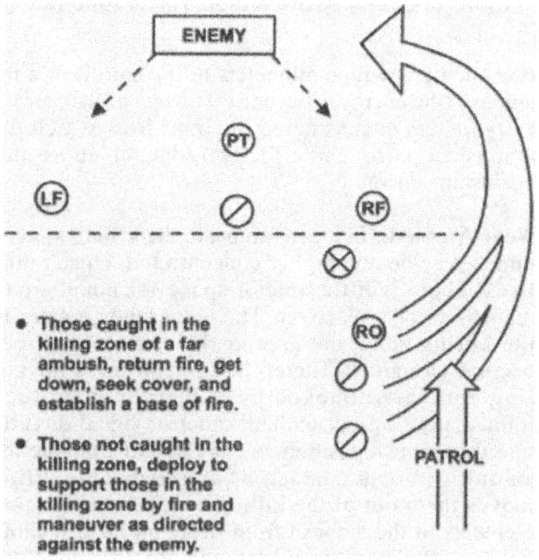

Figure 11-6. Far Ambush.

CHAPTER 12. RECONNAISSANCE PATROLS

The commander needs accurate, timely information about the enemy and the terrain the enemy controls to assist in making tactical decisions. Reconnaissance patrols are one of the most reliable means for obtaining this information. These patrols engage the enemy only when necessary to accomplish their mission or for protection. They generally avoid combat, accomplish the mission by stealth, and do not maintain contact with the unit that sent them out. A reconnaissance patrol is capable of carrying the search for information into the area occupied by enemy forces—usually beyond the range of vision of friendly ground observation posts—and examining objects and events at close range.

12001. GENERAL MISSIONS

General missions for reconnaissance patrols include gathering information about location and characteristics of friendly or hostile positions and installations, terrain (routes, stream crossings), and obstacles.

12002. SPECIFIC MISSIONS

Locate the Enemy

Reconnaissance patrols try to determine the—

- Location of enemy forces, installations, and equipment.
- Identification of enemy units and equipment.
- Strength of enemy forces.
- Disposition of enemy forces.
- Movement of enemy personnel and equipment.
- New or special types of weapons.
- Presence of mechanized units.
- Unusual enemy activity.
- Presence of NBC equipment.

Reconnaissance of Enemy Wire Obstacles

A patrol with the mission of investigating enemy-emplaced wire obstacles employs a formation providing for all-around security and takes precautions against being observed by the enemy. The patrol leader and one Marine inspect each gap and establish its location by means of compass bearings to prominent objects in the rear of enemy or friendly positions or through the use of global positioning satellite receivers.

Reconnaissance of Contaminated Areas

A patrol with the mission of investigating a contaminated area reconnoiters and marks the area's boundaries. Patrol members wear protective clothing and field protective masks. The patrol leader reports the extent of the area, the ttype of agent used, the terrain and vegetation and the method of marking the contaminated area. A sketch of the contaminated area, prepared by the patrol leader, should be included in the patrol report.

Reconnaissance of Enemy Minefields

Patrols assigned to reconnoiter enemy minefields are often composed of specially trained personnel. The type of mine is identified and the possibility of bypassing the area determined.

Reconnaissance of Terrain

The reconnaissance patrol is particularly suited for gathering information about the terrain within their area of operations. (Refer to MCRP 2-15.3B, *Reconnaissance Reports Guide*.) Commanders must know the location and condition of bridges, streams, and roads in order to make proper plans for the movement of troops. The following information should be obtained by a reconnaissance patrol assigned such missions:

- Bridges:
  - Maximum load capacity.
  - Material used in construction (wood, stone, concrete or steel).

  - Material used for abutments (dirt, wood, masonry, concrete).
  - Type and number of supporting members of the bridge.
  - Condition and dimensions of bridge flooring.
  - Presence or absence of mines and demolition charges on the bridge and at entrances/exits.
  - Terrain crossed (ravine, stream, road).
  - Primary use (rail, auto, foot).
  - Location and type of an alternate crossing if bridge is unusable.
- Streams and fords:
  - Width and depth of stream.
  - Composition of the bottom (mud, sand, gravel, rock).
  - Speed of current in miles per hour.
  - Composition and gradient of banks.
  - Surrounding terrain and vegetation cover.
- Roads:
  - Width.
  - Composition of surface.
  - Condition.
  - Road blocks.
  - Gradient/degrees of slopes.
  - Curves (location and length).
  - Pitch and culvert locations.
  - Primary use and frequency of use.

12003. TYPES OF RECONNAISSANCE

Route

Route reconnaissance is a directed effort to obtain detailed information of a specified route and all terrain from which the enemy could influence movement along that route. (Refer to MCRP 5-12C.) It is also described as a form of reconnaissance focused along a specific line of communications—such as a road, railway, or waterway—to provide new or updated information on route conditions and activities along the route. (Refer to MCRP 5-12A, *Operational Terms and Graphics*.)

A route reconnaissance normally precedes the movement of forces. It provides detailed information about a specific route and the surrounding terrain that could be used to influence movement along that route. Considerations include trafficablility, danger areas, critical points, vehicle weight and size limitations, and locations for friendly obstacle emplacement.

The objective area is normally defined by a line of departure, a route, and a limit of advance.

Area

An area reconnaissance is a directed effort to obtain detailed information concerning the terrain or enemy activity within a prescribed area such as a town, ridge line woods or other features critical to operations. (Refer to MCRP 5-12C.) An area reconnaissance could also be made of a single point, such as a bridge or installation. (Refer to MCRP 5-12A.)

Emphasis is placed on reaching the area without being detected. Enemy situations encountered en route are developed only enough to allow the reconnoitering unit to report and bypass.

Zone

A zone reconnaissance is a directed effort to obtain detailed information concerning all routes, obstacles (to include chemical or radiological contamination), terrain, and enemy forces within a zone defined by boundaries. A zone reconnaissance is normally assigned when the enemy situation is vague or when information concerning cross-country trafficability is desired. (Refer to MCRP 5-12C.)

The commander specifies specific routes or areas of interest within the zone. The zone to be reconnoitered usually is defined by a line of departure, lateral boundaries, and a limit of advance.

12004. TASK ORGANIZATION

A reconnaissance patrol is organized around the current structure of the Marine rifle squad with one or more of the squad's fire teams assigned as the reconnaissance element to reconnoiter or maintain surveillance over the objective. At least one fire team acts as a security element whose functions are to—

- Secure the objective rally point (see para. 11003).
- Give early warning of enemy approach.
- Protect the reconnaissance unit.

Reinforcing the squad is based on METT-T. If two squads are required, then the mission should be assigned to a platoon.

A company may send a reconnaissance patrol to specifically reconnoiter a given area (a riverbed or a bridge), or to maintain surveillance on a particular streambed for the next four nights. Reconnaissance patrols perform three basic functions en route to and from the objective:

- Provide control.
- Provide security while moving and in the objective area.
- Conduct reconnaissance or surveillance.

Depending on the size of the reconnaissance patrol, an element must be tasked to reconnoiter the area or zone. In the case of a surveillance mission, two teams are recommended to share the same task (so one could lay up and rest, while the other lays low and hides to maintain strict silence and no movement while surveillance is being maintained on the objective). Whatever the case, each unit that has a surveillance mission or reconnaissance mission must be thoroughly briefed as to what essential elements of information are to be collected as well as their location in the objective area.

The patrol should be organized with one or two fire teams to actually conduct the reconnaissance mission and the remaining fire team to provide security. Each fire team should be prepared to assume either mission. A small area reconnaissance patrol needs only one fire team for the assigned mission. A patrol with a wider area reconnaissance mission should use two fire teams to physically conduct the mission and one fire team for cover and/or security.

The security element for a reconnaissance patrol should be organized to cover the likely avenue of approach into the objective area, to protect the units conducting the reconnaissance, and to cover the objective rally point.

The variations of the special organization for reconnaissance patrols illustrate the patrol leader's flexibility in organizing the patrol to meet mission requirements. The patrol leader is not limited to the variations reflected herein but can choose any special organization as long as that task organization accommodates the requirement for command and control, reconnaissance, and security.

12005. SIZE OF RECONNAISSANCE PATROLS

A reconnaissance patrol should be kept to the minimum number of personnel required to accomplish the mission. A mission requiring a patrol to remain away from its unit for a considerable period of time, or one requiring a patrol to send back information by messenger, increases the size of the patrol. Reconnaissance patrols seldom exceed a squad in strength. Unit integrity should be preserved whenever possible. Intelligence personnel, interpreters, and other specialists, such as radio operators or engineers, are assigned to a patrol if the particular mission requires.

12006. RECONNAISSANCE EQUIPMENT

Patrol members are armed and equipped as necessary for accomplishing the mission. The automatic rifle in each fire team provides a degree of sustained firepower in case of enemy contact. The patrol should have at least two: pairs of binoculars, pairs of wire cutters, maps, compasses, and watches. Night observation equipment may be used. Pencils and small notebooks are carried so notes and sketches can be made. A message book with message blanks and overlay paper is mandatory.

12007. RECONNAISSANCE PATROL ACTIONS AT THE OBJECTIVE AREA

Route

The patrol leader halts and conceals the patrol near the objective area in the final preparation position. The patrol leader then conducts leader's reconnaissance to confirm the plan for positioning the security teams and employing units assigned to the reconnaissance mission. The patrol leader returns to the patrol and positions the security to provide early warning of enemy approach and secure the objective rally point. The reconnaissance unit(s) then reconnoiters the objective area (route). The reconnaissance unit may move to several positions, along or adjacent to the specific route, in order to conduct a thorough reconnaissance. After completing the reconnaissance, each reconnaissance team moves to the objective rally point and reports to the patrol leader. The patrol then returns to friendly lines and the leader makes a full report.

Area

The patrol leader halts and conceals the patrol near the objective area in the final preparation position. The patrol leader then conducts leader's reconnaissance to pinpoint the objective and confirm the plan for positioning the security teams and employing units assigned the reconnaissance mission. The patrol leader returns to the patrol and positions security to provide early warning of enemy approach and secure the objective rally point. The reconnaissance unit(s) then reconnoiters the objective area. The reconnaissance unit may move to several positions, perhaps making a circle around the objective area, in order to conduct a thorough reconnaissance. When the reconnaissance is completed, the patrol leader assembles the patrol and tells members what has been observed and heard. Other patrol members contribute observations. The patrol then returns to friendly lines and the patrol leader makes a full report.

Zone

The patrol leader halts the patrol at the final preparation position, confirms the plan, and conducts leader's reconnaissance. The patrol leader positions the security team and sends out the reconnaissance team. When the entire patrol is used to reconnoiter the zone, it provides its own security. After completing the reconnaissance, each reconnaissance team moves to the objective rally point and reports to the patrol leader. The patrol then returns to friendly lines.

CHAPTER 13. COMBAT PATROLS

Combat patrols are assigned missions that usually include engaging the enemy. They are fighting patrols. Every combat patrol has a secondary mission: gaining information about the enemy and terrain. Combat patrols are employed in both offensive and defensive combat operations and they assist the parent unit in accomplishing its mission by inflicting damage on the enemy; establishing and/or maintaining contact with friendly and enemy forces; denying the enemy access to key terrain; and probing enemy positions to determine the nature and extent of enemy presence.

13001. TASK ORGANIZATION

A combat patrol is organized around the current structure of the Marine rifle platoon. A combat patrol leader should use the unit's normal organization (fire team, squad, and platoon) in assigning functions, patrol missions, and chain of command. Combat patrols must be able to perform the following four basic functions en route to and from the objective:

- Provide control.
- Provide security.
- Provide support by fire.
- Attack or assault the objective.

A rifle platoon could task-organize as a combat patrol as follows:

- Platoon headquarters (patrol headquarters).
- First squad (security).
- Second squad (support).
- Third squad (assault).

Every combat patrol must—

- Provide a control mechanism in the form of a headquarters.
- Designate a unit (a fire team or squad) to provide security while moving en route to the objective or while at the objective. At the objective area, this unit isolates the objective area, secures the objective rally point, and covers the withdrawal.
- Designate a unit to act as support. This unit provides the base of fire in the attack or covers withdrawals or advances.
- Designate a unit(s) to conduct the attack or assault. This unit(s) engages the enemy at the objective area by fire and maneuver or movement. It also operates immediately in the objective area (searching, demolition, prisoners of war, etc.). Paragraph 9001 outlines the general organization of combat patrols. As in the case with reconnaissance patrols, the task organization of a combat patrol depends on the specific mission assigned. If any special requirements are generated because of the specific mission, the patrol is task-organized to fit the needs of the mission.

13002. EQUIPMENT

Combat patrols are armed and equipped as necessary for accomplishing the mission. In addition to binoculars, wire cutters, compasses, and other equipment generally common to all patrols, it usually carries a high proportion of automatic weapons and grenades. Communications with higher headquarters is important as success of the mission may depend on being able to call for supporting fires. Also, internal radio communications with the units and teams may be useful. However, the patrol must not be so overburdened with equipment as to impede movement or mission accomplishment.

13003. RAID PATROLS

A raid is a surprise attack on an enemy force or installation with the attacking force withdrawing after accomplishing its mission. Raids destroy or capture enemy personnel or equipment, destroy installations, or free friendly personnel who have been captured by the enemy. Patrolling techniques are used in planning and when moveing to and from the objective. (Refer to MCWP 3-41.2, *Raids*.) Surprise, firepower, and violence of action are the keys to a successful raid. Patrols achieve surprise by attacking—

- When the enemy is least prepared (e.g., during periods of poor visibility such as darkness, rain, fog, or snow).
- From an unexpected direction. (This might be accomplished by approaching through a swamp or other seemingly impassable terrain.)
- With concentration of firepower at critical points within the objective.

Planning

A successful raid requires detailed planning. The leader of a combat patrol engaged in raiding must anticipate probable situations and decide upon definite courses of action to meet them. Rehearsals are imperative.

A raid patrol conducts such missions as destroying an enemy outpost or seizing prisoners from an observation post or lightly defended position.

While preparing for the mission, the patrol leader requests fire support required for the accomplishment of the mission. If practical, artillery and mortars should be employed to isolate the objective to prevent movement of enemy reinforcements into the area.

Execution

The leader's plan must be detailed and complete. All of the considerations outlined in chapter 11 must be covered. Patrol formations must provide for ease of control and all-around security while moving to and from the objective area and provide for rapid and coordinated deployment of the various units once the objective area is reached. The leader's plan usually includes the encirclement of the hostile position—either physically or by fire—in order to isolate it during the assault.

The final simultaneous assault against the objective develops when enemy defensive fires at the objective are suppressed by either friendly fire superiority or surprise. The assault is covered by the fire of the unit assigned the function of support by fire.

Grenades, SMAWs, and demolitions are most effective for clearing bunkers.

Security units are posted to isolate the objective. The patrol leader signals them when the withdrawal begins. As a minimum, security is on each flank and to the rear (at the objective rally point).

Actions at the Objective

The patrol leader halts the patrol near the objective at the final preparation position. Security is established and the leader's reconnaissance is made with appropriate subordinate leaders. When the leaders return to the patrol, they confirm previous plans or announce any changes. Movements are arranged so all units reach their positions simultaneously. This improves the patrol's capability for decisive action, if prematurely detected by the enemy.

The teams of the security element move to positions to secure the objective rally point, give early warning of enemy approach, block avenues of approach into—and prevent enemy escape from—the objective area. As the assault element moves into position, the security element informs the patrol leader of all enemy activity, firing only if detected or on the patrol leader's order. Once the assault element commences action, the security element prevents enemy entry into or escape from the objective area. The security element covers the withdrawal of the assault element (and support element, if employed) to the objective rally point, withdrawing only upon order or prearranged signal.

As the assault element approaches the objective, it deploys early enough to permit immediate assault if detected by the enemy. Each team uses stealth while moving into proper position. On command, or if one or more of the assault element is detected and fired upon by the enemy, the support element opens fire to neutralize the objective, then ceases or shifts fire according to prearranged plans and signals. As supporting fires cease or shift, the assault element assaults the objective. Demolition, search, and other teams are protected by the assault element while they work. On order, the assault element withdraws to the objective rally point.

If a support element is employed, its leader deploys teams to provide fire support for the assault element. Each member of the support element must know the scheme of maneuver to be used by the assault element, specific targets or areas to be neutralized by fire, and the signals that will be employed to commence, shift, and cease fires. The support element withdraws on order of the patrol leader. At the objective rally point, the patrol leader quickly reorganizes the patrol and begins the return movement to friendly lines.

13004. CONTACT PATROLS

General

Contact patrols establish and/or maintain contact to the front, flanks or rear by—

- Establishing contact with an enemy force when the definite location of the force is unknown.
- Maintaining contact with enemy forces through direct and/or indirect fires, or observation.
- Avoiding decisive engagement with the enemy.

Task Organization and Equipment

Task organization and equipment depend on the known enemy situation and anticipated enemy contact. A patrol sent out to establish contact with an enemy force is organized, armed, and equipped to overcome resistance of light screening forces in order to gain contact with the main enemy force. It is not organized and equipped to engage the main enemy forces in combat. Communication is paramount; radios must be reliable over the entire distance covered.

Actions at the Objective

The patrol leader selects a series of objectives. Once an objective is reached, the patrol leader initiates a planned set of actions in order to establish and maintain contact with the enemy. These plans and actions are guided by the missions to establish or maintain contact—not to engage in decisive combat. Contact with the enemy is maintained for the purposes of surveillance, applying pressure, and preventing seizure of the initiative. If the contact patrol becomes decisively engaged with the enemy, many of the tasks originally assigned to the patrol cannot be accomplished, since the enemy has seized the initiative and friendly forces are not forced to react.

13005. AMBUSH PATROLS

General

An ambush is a surprise attack from a concealed position upon a moving or temporarily halted target. It is one of the oldest and most effective types of military actions. Ambush patrols conduct ambushes of enemy patrols, resupply columns, and convoys. The intent of an ambush is to place the enemy in a dilemma where staying in the kill zone or attempting to move out of it prove equally lethal. The ambush may include an assault to close with and decisively engage the enemy, or the attack may be by fire only.

Purpose of Ambushes

Ambushes are executed for the general purpose of reducing the enemy's over-all combat effectiveness and for the specific purpose of destroying its units. The cumulative effect of many small ambushes on enemy units lowers enemy troop morale and harasses the enemy force as a whole.

Destruction is the primary purpose of an ambush because loss of men killed or captured, and loss of equipment and supplies destroyed or captured, reduces the overall combat effectiveness of the enemy.

Harassment, though less apparent than physical damage, is a secondary purpose of ambushes. Frequent ambushes force the enemy to divert men from other missions to guard convoys, troop movements, and carrying parties. When enemy patrols fail to accomplish their mission because they are ambushed, the enemy is deprived of valuable information. A series of successful ambushes causes the enemy to be less aggressive and more defensive minded. His men become apprehensive, overly cautious, reluctant to go on patrols, seek to avoid night operations, are more subject to confusion and panic if ambushed, and in general, decline in effectiveness.

Classification of Ambushes

A *deliberate ambush* is one in which prior information about the enemy permits detailed planning before the patrol departs for the ambush site. Information needed to plan a deliberate ambush includes the size, composition, and organization of the force to be ambushed; how the force operates; and the time it will pass certain points or areas. A deliberate ambush may be planned for such targets as—

- Any force if sufficient prior information is known.
- Enemy patrols that establish patterns by frequent use of the same routes or habitually depart and re-enter their own areas at the same point.
- Logistic columns.
- Troop movements.

An *ambush of opportunity* is conducted when available information does not permit detailed planning before the patrol departs. This is the type of ambush that an infantry unit normally conducts. An ambush of opportunity should not be confused with a hasty ambush. An ambush of opportunity is a planned ambush; a hasty ambush is an immediate action. In planning for an ambush of opportunity, the patrol must be prepared to execute any of several courses of action based on the types of targets that may be ambushed and must rehearse prior to departure. The course of action taken is determined when the opportunity for ambush arises.

The patrol leader may be directed to reconnoiter an area for a suitable ambush site, set up at the site selected, and execute an ambush against the first profitable target that appears.

The patrol may depart just after dark, move to a specific point, observe until a designated time, ambush the first profitable target after that time, and return before daylight.

A *hasty ambush* is an immediate action where the patrol makes visual contact with an enemy force and has time to establish an ambush without being detected. The actions for a hasty ambush must be well rehearsed and accomplished through the use of hand and arm signals given from the patrol leader.

Types of Ambushes

There are two types of ambushes: point and area. The *point ambush* is one where forces are deployed to attack along a single killing zone. The *area ambush* is one where forces are deployed as multiple related point ambushes.

A point ambush, whether independent or part of an area ambush, is positioned along the enemy's expected route of approach. Formation of the forces conducting the ambush is important because, to a great extent, the formation determines whether a point ambush is able to deliver the heavy volume of highly concentrated fire necessary to isolate, trap, and destroy the enemy.

The ambush formation to be used is determined by careful consideration of possible formations and the advantages and disadvantages of each in relation to—

- Terrain.
- Visibility.
- Forces.
- Weapons and equipment.
- Ease or difficulty of control.
- Target to be attacked.
- Combat situation.

For a detailed discussion of ambush formation, see appendix D.

Ambush Operation Terms

The *ambush site* is the location where an ambush is established.

The *killing zone* is that portion of an ambush site where fires are concentrated to trap, isolate, and destroy the target. On little-traveled roads, an obstacle placed in a defile, in the woods, on a bridge or on a steep upgrade can be used effectively to force vehicles to halt, and thus render the occupants vulnerable to attack. Antitank mines may be emplaced and the occupants of the wrecked vehicle killed or captured while still dazed by the explosion.

A *near ambush* is a point ambush where the attacking force is located within reasonable assaulting distance of the killing zone (50 meters is a guide). A near ambush is most often conducted in close terrain, such as a jungle or heavy woods.

A *far ambush* is a point ambush where the attack force is located beyond reasonable assaulting distance of the killing zone (beyond 50 meters is a guide). A far ambush may be more appropriate in open terrain offering good fields of fire or when the target will be attacked by fire only.

Factors for a Successful Ambush Patrol

There are many factors that give the ambush its best chance of success. The ideal situation would be to position the ambush on favorable terrain and have detailed planning completed beforehand.

The patrol leader of an ambush looks for the most favorable terrain in which the enemy is canalized between two obstacles with limited opportunity to attack or escape. Suitable areas include defiles, small clearings, bends in trails, and steep grades. Dense undergrowth adjacent to the ambush site permits observation from concealed positions. The ambush patrol should have maximum cover and concealment, not only for the firing positions, but for the routes of withdrawal. The enemy should be in an area offering as little protection from fire as possible. Favorable fields of fire include stretches of road, trail or open ground of at least 100 meters for machine guns and 15 meters for rifle fire and grenades. The ambush site can be improved by constructing obstacles—such as felled trees, wire, land mines, or booby traps—to impede the enemy.

Planning

A deliberate ambush or an ambush of opportunity requires thorough planning.

A deliberate ambush plan is based on extensive knowledge of the enemy and terrain, and is planned and rehearsed in great detail. A physical reconnaissance of the ambush site is made during the preparation phase and information gained is incorporated into the plan. All likely immediate actions of the enemy when ambushed are examined. Planned counteractions are developed and rehearsed.

In planning an ambush of opportunity, any available information on the enemy and terrain is used. A tentative plan for the ambush that incorporates all anticipated actions is developed and rehearsed. However, the bulk of planning is done concurrently during the patrol leader's reconnaissance of the prospective ambush site. In a rapidly developing situation, hasty ambush immediate action is employed.

The route and ambush site considerations apply to both deliberate ambushes and ambushes of opportunity. A primary route that allows the patrol to enter the ambush site from the rear is planned. Entering the prospective killing zone is avoided. If the killing zone must be entered to place mines or explosives, care is taken to remove any tracks and signs that might alert the enemy and compromise the ambush. If mines or explosives are to be placed on the far side of the ambush site, or if the appearance of the site from the enemy's viewpoint is to be checked, a wide detour is made around the killing zone. Care is taken to remove any tracks that might reveal the ambush. An alternate route from the ambush site to the objective rally point, as in other patrols, is planned.

Maps and aerial photographs are used to carefully analyze the terrain. When possible, an on-the-ground reconnaissance of the ambush site is made prior to occupation. Obvious ambush sites are avoided as the element of surprise is even more difficult to achieve in these areas. An ambush site must provide for—

- Favorable fields of fire.
- Occupation and preparation of concealed positions.
- Canalization of the target into the killing zone. (An ideal killing zone restricts the enemy on all sides, confining him to an area where he can be quickly and completely destroyed. Natural obstacles, such as cliffs, streams, embankments, or steep grades, are used whenever possible to force vehicles to slow down. Man-made obstacles, such as barbed wire, mines, and craters in the roads, are used to supplement natural obstacles.)
- Covered routes of withdrawal that enable the ambush force to break contact.
- Avoidance of enemy pursuit by fire.

Occupation of Ambush Site

The surrounding area is searched for enemy patrols prior to occupation of the ambush site. Ambush formations are used to physically deploy the patrol in a manner to inflict maximum destruction upon the enemy and to provide maximum security to the patrol. Ambush formations are contained in appendix D.

Positions

The patrol is moved into the ambush site from the objective rally point. Security is positioned first to prevent surprise while the ambush is being established. Automatic weapons are then positioned so each can fire along the entire killing zone. If this is not possible, overlapping sectors of fire are provided to cover the entire killing zone. The patrol leader then selects his position, located so he can tell when to initiate the ambush. Riflemen and grenadiers are positioned and sectors of fire are assigned to cover any dead space left by the automatic weapons. The patrol leader sets a time by which positions are to be prepared. Patrol members clear fields of fire and prepare positions in that order, with attention to camouflage for both.

Suitable Objective Rally Point

An easily located objective rally point is selected and made known to all patrol members. The objective rally point is located far enough from the ambush site so that it will not be overrun if the enemy assaults the ambush. Routes of withdrawal to the objective rally point are reconnoitered. Situation permitting, each man walks the route he is to use and picks out checkpoints. When the ambush is to be executed at night, each man must be able to follow his route in the dark. After the ambush has been executed, and the search of the killing zone completed, the patrol is withdrawn quickly but quietly, on signal, to the objective rally point where it reorganizes for the return march. If the ambush was not successful and the patrol is pursued, withdrawal may be by bounds. The last group may arm mines, previously placed along the withdrawal route, to further delay pursuit.

Local Security

Security must be maintained. Security elements do not usually participate in the initial attack, but protect the rear and flanks, and cover the withdrawal.

Patience

The Marines of the ambush force must control themselves so that the ambush is not compromised. Patience and self-discipline are exercised by remaining still and quiet while waiting for the target to appear, particularly if the patrol occupies the ambush site well ahead of the arrival of the enemy. Patience is necessary so as not to alert the enemy to the presence of the ambush.

Surprise

Surprise must be achieved, or the attack is not an ambush. If complete surprise cannot be achieved, it must be so nearly complete that the target is not aware of the ambush until too late for effective reaction. Surprise is achieved by careful planning, preparation, and execution so that targets are attacked when, where, and in a way for which they are least prepared.

Coordinated Fires

Properly timed and delivered fires contribute heavily to the achievement of surprise, as well as to destruction of the enemy. The lifting or shifting of fires must be equally precise; otherwise, the assault is delayed and the enemy has an opportunity to recover and react. All weapons, mines, and demolitions are positioned and all fires, including those of available artillery and mortars, are coordinated to achieve the following results:

- Isolation of the killing zone to prevent the enemy's escape or reinforcement.
- Surprise delivery of a large volume of highly concentrated fires into the killing zone.

Control

Close control of the patrol is maintained during movement to, occupation of, and withdrawal from the ambush site. This is best achieved through rehearsals and establishment and maintenance of good communications. When the enemy approaches, the temptation to open fire before the signal is given is resisted. The patrol leader must effectively control all elements of the ambush force. Control is most critical at the time the enemy approaches the killing zone. Control measures must provide for—

- Early warning of enemy approach.
- Fire control. Withhold fire until the enemy has moved into the killing zone, then open fire at the proper time.
- Initiation of appropriate action, if the ambush is prematurely detected. Individual patrol members must be prepared to react if detected by the enemy prior to the initiation of the ambush.
- Timely and orderly withdrawal of the ambush force from the ambush site and movement to the objective rally point.

It is important to remember that an ambush patrol should have four distinct signals: one to open fire (with an alternate signal to open fire to be used at the same time as the primary); a signal to cease fire or shift fire; a signal to assault or search the killing zone; and a signal to withdraw. The signal to open fire should meet two criteria: first, it should be the firing of a weapon that will kill the enemy; secondly, it should be a weapon reliable in any weather condition. A good primary signal is a Claymore mine, and an alternate signal would be a closed bolt weapon (M16A2). Open bolt weapons (M240G, M249) should not be relied upon to initiate an ambush.

Execution of an Ambush

The manner in which the patrol executes an ambush depends primarily on whether the ambush's purpose is harassment or destruction. To a lesser degree, the execution of the ambush is determined by whether the ambush is deliberate or an ambush of opportunity.

When the primary purpose is harassment, the patrol seals off the area with security teams to prevent enemy reinforcement and escape. Maximum damage is inflicted with demolitions and automatic weapons fire. The patrol delivers a very heavy volume of fire for a short time and withdraws quickly and quietly. The patrol avoids being seen by the enemy.

When the primary purpose of the patrol is destruction, the area is sealed off with security units. Maximum damage is inflicted with demolitions, antitank weapons, and automatic weapons fire from the support team or element. When these fires cease or shift, an assault is launched into the killing zone with heavy fire and violence to complete destruction. The assault unit provides security, while designated teams search and/or capture personnel and destroy vehicles and equipment. On the patrol leader's command, or by prearranged signal, all units withdraw to the objective rally point and move out quickly.

When the patrol's primary purpose is to obtain supplies or capture equipment, security units seal off the area. Demolitions and weapons are used to disable vehicles. The assault unit must use care to ensure its fire does not damage the desired supplies or equipment. Designated teams secure the desired items; other teams then destroy enemy vehicles and equipment.

The most successful ambush is one where the attacker is deployed and concealed in such a way that the enemy will unknowingly be surrounded by fire. The usual method is for the attackers to deploy themselves along a trail or route the enemy will travel. The enemy is permitted to pass by the center of the attacker's force so that the attack can be made from the front, flank, and/or rear. One or two men are posted well forward and to the rear along the route to prevent any enemy from escaping. All fires should be delivered simultaneously on a prearranged signal.

An effective method of luring the enemy is for an ambushing patrol to cut communication or electrical wire. The patrol then deploys and ambushes the line repair crew when it arrives. Since the line crew may be protected by riflemen, the attackers must be careful to engage the entire party.

Vehicles and foot personnel moving on well-established transportation routes can sometimes be captured by altering or moving directional signs so as to divert the enemy into an area where he can be more readily attacked. The attack can best be accomplished at an obstacle, such as a stream or gully, that forces the enemy to stop or slow down.

After the enemy has been ambushed and destroyed, the unit quickly withdraws over a prearranged route to the objective rally point. Speed is very important, since the noise of the ambush could alert other nearby enemy units.

13006. SECURITY PATROLS

General

Security patrols are assigned missions that may or may not require them to engage the enemy. They are used in proximity to defensive positions, on the flanks of advancing units or in rear areas. Purposes of security patrols are to detect infiltration by the enemy, destroy infiltrators, and protect against surprise and ambush.

In any situation where there is a threat of attack, such as a rear area threatened by guerrillas or a facility that is under threat of a terrorist attack, all Marines, not just the infantry, must know how to conduct a security patrol.

In just the offensive operations, infantry units provide security patrols to screen their flanks, areas, and routes. Whereas, in defensive operations, security patrols are used to prevent the enemy from infiltrating an area, detect and destroy infiltrators, and prevent surprise attacks. In rear areas, particularly when there is guerrilla or terrorist threat, the requirement to conduct security patrols increases for all Marine air-ground task force (MAGTF) units ashore, particularly aviation and combat service support units.

Task Organization and Equipment

Generally, a Marine rifle squad or similar sized organization is considered ideal for security patrols. Communications are important to higher headquarters so that they receive information from the patrol; and communications are important to the patrol to request fire support, etc. The radio the patrol carries must have the range necessary for higher headquarters to be able to receive transmissions from anywhere along the patrol route, and the patrol must have a secondary means of reporting (i.e., flare signals upon contact).

Patrol Procedures

All of the procedures presented in previous sections are to be used in security patrols.

Patrol Planning

Security patrol planning includes—

- Rehearsing prior to departing friendly lines.
- Maintaining communications.
- Support by organic weapons.
- Reinforced if necessary.
- Using varied routes and never establishing a routine pattern.
- Staying within proximity of friendly units.

Patrol Techniques

Within rear areas, an irregular pattern of patrol is established and changed daily. Outside of friendly lines it would be prudent to establish a definite preplanned route for the patrol, of which all adjacent units know the route. The parent unit commander establishes frequent checkpoints for control. If checkpoints are designated, the patrol leader treats them as individual objectives to be searched and cleared.

The patrol has a definite plan as to what to do if contact with enemy is made, how to break contact, how to defend itself, and how to call for supporting fires. It is imperative that patrol members know what to do if they become split or separated; i.e., location of rally points and how to be recovered.

13007. URBAN PATROLS

General

As national strategy continues to focus on regional vice global conflicts, the Marine Corps will continue to conduct urban operations in various operational environments. Cities and towns are often the center of economic and political power and are therefore extremely vulnerable to urban insurgent activities and violence. The lessons learned from recent operations in Somalia, as well as experiences gained by British forces in Northern Ireland, provide the foundations of urban patrolling.

Enemy Forces

Enemy forces in urbanized areas range from organized military forces to low intensity engagements with insurgents, such as terrorists or local gangs. For Marine Corps doctrine and tactics, techniques, and procedures (TTP) for dealing with the higher intensity threat in urbanized areas, refer to MCWP 3-35.3, *Military Operations on Urbanized Terrain.*

Common Threat Tactics

Cities provide cover and concealment for both friendly forces and enemy forces. However, enemy forces generally find active support only in certain areas of the town or city. The urban-based insurgent or terrorist usually lives in a friendly community or in one where the people are too frightened to withhold support or inform anyone about the situation. The insurgent or terrorist normally maintains close contact with leaders and others friendly to the cause. The enemy will often have efficient communication and intelligence systems, sometimes involving women and children to provide cover for its activities.

The urban insurgent normally cannot, like his rural counterpart, establish bases and recruit large military units. He is generally an individual or a member of a relatively small group. He relies on the cover afforded by the city's populace and terrorizing them to coerce loyalty or support.

Urbanized areas tend to give the insurgent and/or terrorist many opportunities to initiate action and gain advantage. The normal presence of large numbers of people in cities provides the foe an opportunity to mass crowds quickly and manipulate demonstrations easily. The presence of women and children during mass demonstrations may restrict the courses of action available to friendly forces. Major incidents stemming from overreaction or excessive use of force by friendly forces may provide the insurgent with propaganda material. Publicity is easily gained in an urbanized area because major incidents can't be completely concealed from the local population. Insurgent successes can be exploited to discredit the ability of host nation police, friendly forces, and the civil government, and gain recognition for the insurgents' cause.

The urban insurgent or terrorist can usually be expected to operate more boldly than his rural counterpart. This is reflected in the enemy's tactics. A single sniper or bomber may be the norm in the urbanized area, whereas the rural threat is generally the more conventional ambush. In urbanized areas, explosive devices can be easily emplaced and used effectively against large groups or select individuals. Enemy forces may be expected to employ the following tactics in urbanized areas:

- Using local communications, such as radio and newspapers, for propaganda purposes.
- Disrupting industry and public services through strikes and sabotage.
- Generating widespread disturbances designed to stretch the resources of the security force.
- Creating incidents or massing crowds in order to lure the patrol or reaction force into a trap.
- Provoking security forces in the hope that they may react improperly, therefore discrediting the security force by means of propaganda.
- Sniping at roadblocks, outposts, sentries, and patrols.
- Attacking friendly bases with rockets and mortars.
- Planting explosive devices, either against specific targets or indiscriminately, to cause confusion and destruction, and lower public morale and confidence.
- Using ambush patrols.
- Firing on friendly helicopters.

Principles of Urban Patrolling

Patrolling in an urban environment often presents conditions considerably different and often more complex than those encountered in rural and less inhabited areas. While the principles of patrolling are still relevant in an urban situation, the nature of urban patrolling has led to the development of six specific urban patrolling principles. They are—

1. **Depth.** The restrictive, canalizing nature of urbanized terrain usually limits a patrol's ability to disperse laterally. To prevent the patrol from bunching up, patrols normally maintain dispersion along the length of a patrol formation.

2. **Mutual Support.** The positioning of units in-depth within the patrol enables one unit to cover another unit's movement and facilitates immediate action during various situations. Aircraft, vehicles, and snipers also provide good mutual support.

3. **Deception and Pattern Avoidance.** Deception and pattern avoidance are normally a planning consideration of the headquarters directing the entire patrolling plan. By varying patrol routes, durations, and departure times, hostile actions commonly used against urban patrols—such as enemy ambushes, and roadblocks—are more difficult to plan and may be preempted.

4. **Intra-patrol Communication.** Elements of an urban patrol must have the means to communicate with each other. Ideally, each element will possess a radio enabling it to remain in continuous communication that facilitates rapid response and reporting to higher headquarters, reaction force coordination, and coordination of actions with other patrols or fire support agencies.

5. **Establishment of a Reaction Force.** The nature of urbanized terrain (its compartmentalization) makes urban patrols more vulnerable to a wider range of hostile actions. The requirement for immediate, coordinated reinforcement of a patrol is best satisfied by employment of an established reaction force. The reaction force requires superior mobility (relative to the enemy's) and fire support to be effective for this task.

6. **Three-Dimensional Threat**. Patrolling in an urban environment requires constant attention to its three-dimensional aspect; hostile actions can originate from rooftops, streets, subsurface levels or combinations of all levels at once.

Classification of Urban Patrols

Mission

The vast majority of urban patrols are overt in nature, with their presence readily apparent to the local populace. Most urban patrols are combat vice reconnaissance patrols. The vast majority of urban patrols are security patrols. However, units may be assigned secondary tasks of reconnoitering specific or general areas along the patrol routes.

Raids normally involve a swift penetration of an objective to secure information, confuse the enemy or destroy installations. Raids include a planned withdrawal upon completion of the assigned mission.

Movement

Means of movement are as follows:

- Dismounted: movement on foot.
- Vehicular: movement by motorized, mechanized or armored vehicle.
- Helicopterborne: movement by helicopter, however, helicopterborne patrols will usually involve dismounted or vehicular movement after patrol insertion.
- Combination: movement using a combination of methods.

Task-Organization

Units task-organize to meet the specific requirements of the mission and situation. The Marine rifle squad is ideally suited for urban patrolling and can easily integrate attached specialists required by specific missions. Specialists who may accompany urban patrols include—

- Interrogator-translator team (ITT) and counterintelligence team (CIT) Marines.
- Explosive ordnance disposal (EOD) personnel.
- Members of host nation or allied military forces.
- Interpreters.
- Local community leaders.
- Local law enforcement officers.
- Public affairs personnel escorting media representatives.

Dismounted Patrol Organization

Patrols should maintain unit integrity (fire team and/or squad) in organizing elements for an urban patrol. Similar to reconnaissance patrols, the vulnerability of urban patrols necessitates that all elements must provide for their own security in addition to the combat aspect of the mission. Combined assault and security teams are an effective method to organize for all-around security.

Urban Patrol Base Operations

Urban patrols may operate from an established patrol base that may be located within the unit's assigned area of operations or an area designated for a patrol. The patrol base should be located in a building used exclusively for this purpose. Patrol bases may also be located within a larger site that houses other agencies, such as a higher headquarters' command echelon.

If located within a larger site, the patrol base will be included in the overall facility security plan. If isolated, the patrol base must consider the following security factors:

- External security:
  - Barrier plan.
  - Sentry posts.
  - Local security patrols.
- Internal security:
  - Covered positions for all Marines.
  - Contingency plan for hostile actions against patrol base.

The Reaction Force

The high probability of an urban patrol becoming involved in a hostile or volatile incident requires the establishment of a dedicated reaction force for rapid reinforcement, support or extraction of the patrol. Ideally, the reaction force is—

- Large enough and task-organized in a manner that it can meet and quickly defeat the expected threat. Reaction forces are normally tiered with a lead element (normally one-third of the unit's size) and a main body force (the other remaining two-thirds of the force). Reaction force response times routinely are determined in advance by higher headquarters.
- Ready to respond immediately.
- Motorized or mechanized and supported by close air support and other fire support.
- Familiar with the area of operations.
- Briefed on the patrol's plans and monitors the status of patrols in progress.
- Task-organized to be multimission capable.
- Able to communicate with the higher headquarters, fire support assets, patrol base, and the patrol.
- Controlled by higher headquarters, once employed.

Patrol Preparation

Planning

Higher headquarters will—

- Designate the area for patrol.
- Provide intelligence briefs and updates.
- Ensure liaison with allied forces and the civilian populace.
- Provide special equipment and personnel required for the mission (scout snipers, public affairs officer, interpreters, etc.)
- Provide urban maps, photos, terrain models as required.
- Consider deception and pattern avoidance when issuing mission.
- Prescribe rules of engagement (ROE).

Intelligence Brief

An intelligence brief is conducted by the S-2 officer or representative prior to a patrol conducting its mission. The brief adresses the situation relevant to the specific patrol (e.g., routes, areas, updated enemy situations).

Coordination

Higher headquarters will effect liaison with adjacent and allied forces, as well as civil authorities and other agencies, having a possible effect on the patrol. The patrolling unit generally follows the same procedures as those used during patrol planning and execution in a jungle or forest environment.

The Urban Patrol Order or Warning Order

The Urban Patrol or Warning Orders use the same format and considerations as noted in this publication for patrolling. They rely heavily on a detailed terrain model, photographs, and subterranean construction to ensure complete understanding of the plan.

Rehearsals

The limited size of the patrol base usually precludes the need for full-scale rehearsals. Immediate action drills, such as crossing danger areas, are rehearsed in as much detail as possible, despite the limited available space.

Inspections

Initial and final inspections are conducted in the same manner as other patrols. Attached personnel must be fully integrated into the patrol and familiar with the plan and unit standing operating procedures (SOPs).

Conducting an Urban Patrol

Movement

Individual and unit movement considerations are generally the same as those for other patrols. However, urban environments require consideration of additional factors. Because of these factors, an urban patrol leader should—

- Ensure that each movement within a patrol takes place under the observation or cover of another individual or element of the patrol.
- Know where cover can be taken in the event of a hostile incident or action.
- Be prepared for contact with civilians, especially children, during the patrol and be aware that they may intentionally attempt to distract patrol members.
- Expect the presence of vehicles (both moving and stationary) along the patrol route.
- Expect members of the patrol to be approached by dogs and what action to take if threatened.

Patrol Formations

Squad-sized Patrols. The need for immediate fire power outweighs the dangers of becoming canalized. In contrast to other types of patrols, the headquarters element of an urban patrol will normally locate at the lead of the patrol column. This allows the patrol leader greater flexibility through control of two combined assault and security (A&S) teams. The leaders of these elements tactically stagger their members on each side of the road (see fig. 13-1 on page 13-12).

A&S teams follow in trace of the headquarters element and maintain unit integrity on separate sides of the street. One unit will remain slightly to the rear to create a staggered interval between Marines on either side of the street. This allows A&S teams to take lateral routes in support of headquarters element without having to cross a street to do so.

Platoon-sized Patrols. Squads will generally travel abreast of each other, moving along parallel routes. The interval between squad-sized units and/or teams is situation-dependent, but is usually between 100 and 150 meters (roughly two city blocks; this often prevents visual contact between the units). The intent is to create less of a target to an aggressor, yet still allow the patrol to quickly react to an incident. Individuals within units or teams will move in a staggered column as in a squad-sized patrol. See figure 13-2, which is located on page 13-13).

Night

Night patrols will generally be at least squad-sized and will generally use the same formation as that for day patrols. At night, it may be necessary to close distances between individuals or elements to maintain control. Consideration should be given to the use of night vision devices and thermal weapons sights.

Navigation, Control, and Security Measures

The patrol leader is ultimately responsible for the navigation. The headquarters element normally functions as the base unit during movement. The designated navigator is normally assigned from within the headquarters element. City maps are often inaccurate or outdated; however, when used with aerial photographs and other navigational aids, they can be effective for urban navigation.

Checkpoints and phase lines should be related to major streets (alleys, buildings, bridges) for easy identification. Arrival at checkpoints and crossing of phase

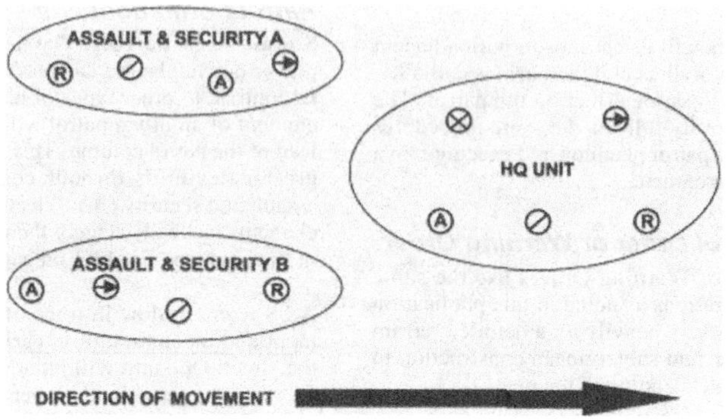

Figure 13-1. Squad-sized Dismounted Urban Patrol Formation.

lines should be relayed to higher headquarters using established brevity or codewords. A detailed patrol overlay is issued to both the reaction force and higher headquarters to keep them advised of routes and control measures used. A terrain model should be maintained at higher headquarters to aid in supervision and control of the patrol.

The use of camouflage should be limited to avoid frightening and confusing the local populace. The often overt nature of urban patrols may negate the need for camouflage. Patrols are sometimes deployed to show force presence and usually move on the urban street in plain view.

During daylight, patrols will routinely vary their rate of movement ranging from short halts to brief periods of double-timing. The British term for this urban patrolling technique is *hard-targeting*, meaning it makes the patrol harder for an enemy to target. Altering the rate of movement is intended to frustrate the enemy's ability to coordinate an attack or ambush against a targeted patrol.

Patrols should use short security halts, with Marines taking up mutually supporting firing positions. Marines must always work in pairs, ensuring mutual support. The last Marine in the element will provide rear security, but stays in his buddy's sight.

Individual Tasks

Individuals may be assigned collateral tasks performed throughout the patrol that may increase the patrol's situational awareness. Individual tasks may include—

- Vehicle spotter: looks for suspicious or known insurgent vehicles.
- Personnel spotter: observes and attempts to recognize previously identified enemy in crowds.
- Talker: attempts to gain information from casual conversations with the local populace. (Talkers are usually subordinate leaders or Marines with foreign language skills.)
- Searcher: conducts physical searches of vehicles and personnel while other patrol members provide cover and security.
- Marksman: engages point targets when the tactical situation does not permit massed or high volume fire. Other patrol members provide security to cover the marksman's engagement.

Departure of Friendly Lines

Urban patrols must vary their departure times to prevent being ambushed while exiting the patrol base.

Individual elements will usually depart exits at staggered times and at different movement rates, especially where sentries cannot provide cover. (This technique is used when the environment adjacent to the patrol base is dangerous.)

Once an element has exited friendly lines, a short halt is conducted in a predetermined, covered initial rally point 50 to 100 meters from the base. This ensures all elements are in position before the patrol continues.

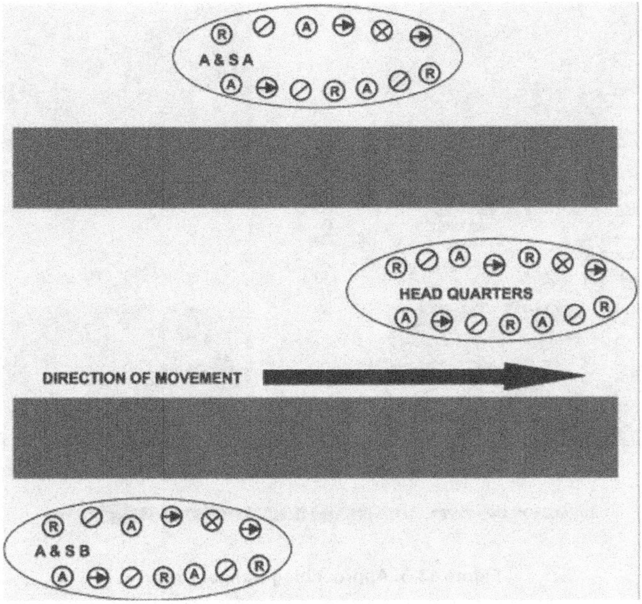

Figure 13-2. Platoon-sized Dismounted Urban Patrol Formations.

Exit points, routes from the base, departure techniques, and locations of IRPs should be varied constantly to avoid setting a pattern. This pattern avoidance may also include using vehicles to insert patrols away from the urban patrol base and employing empty vehicles as part of a deception plan.

Danger Areas

Urban patrols may encounter hundreds of danger areas during a single patrol. The three-dimensional threat requires keen situation awareness by every patrol member. Many danger areas can be dealt with simply by avoidance, while others require an adjustment of patrol formation, movement rate, etc. In the urban environment, places to be treated as danger areas are points that pose a major threat to the patrol, such as local political and religious headquarters, weapons containment areas, roads and routes that canalize movement and direct fire, and any area with a history of repeated contact.

Near and far side rally points are designated and briefed during the issuance of the patrol order. Squad-sized formations may use the A&S teams to provide flank security for the headquarters element and for each other. The headquarters element identifies the danger area and takes up a position on the near side of intersection. Individuals provide all-around security (see fig. 13-3 on page 13-14).

Two Marines (one from each A&S team) are designated to move through the headquarters element and establish respective firing positions on the near side of the danger area covering the patrol's near side flanks. They are followed by a second pair (again, one Marine from each A&S team) that moves across to the far side of the danger area and establishes respective firing positions covering the patrol's far side flanks (see fig. 13-4 on page 13-14).

Once near and far side flank security is established, the headquarters element moves across to the far side of danger area (see fig. 13-5 on page 13-15).

The remaining A&S team members then cross the danger area and join the headquarters element on the far side (see fig. 13-6 on page 13-15).

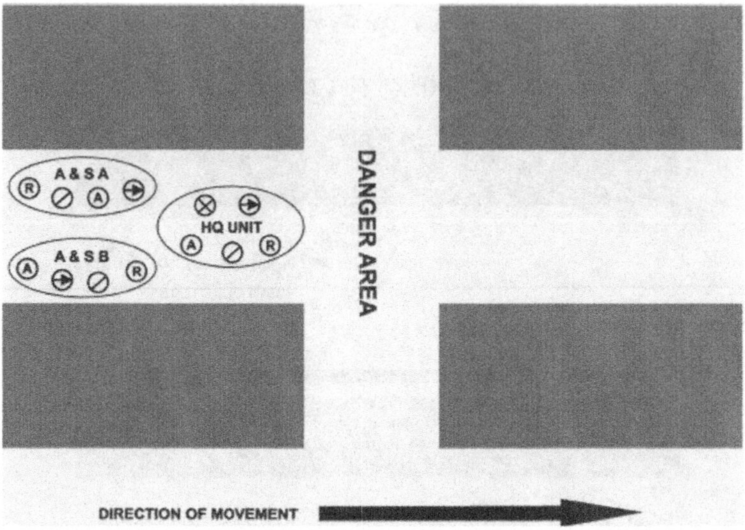

Figure 13-3. Approaching Danger Area.

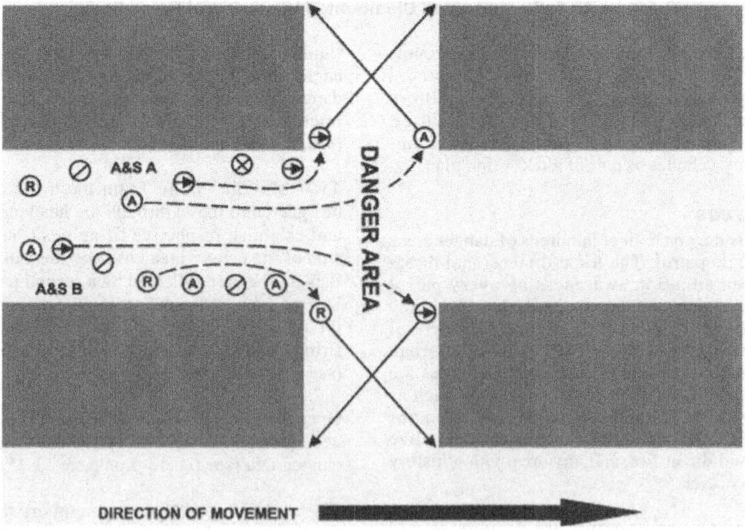

Figure 13-4. Securing Flanks.

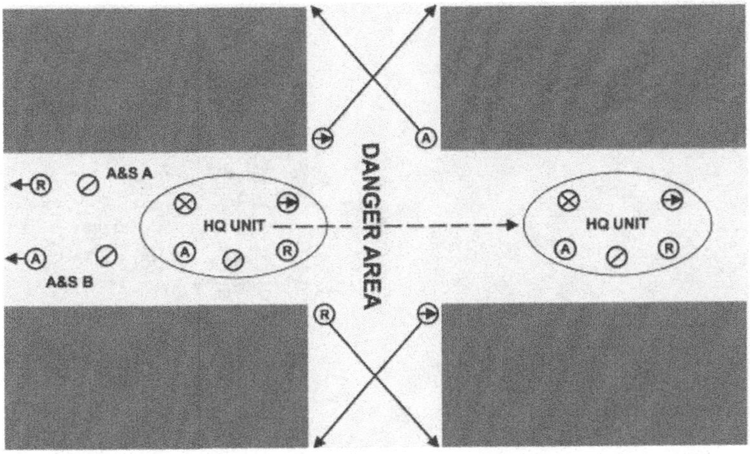

Figure 13-5. HQ Element Moves Through.

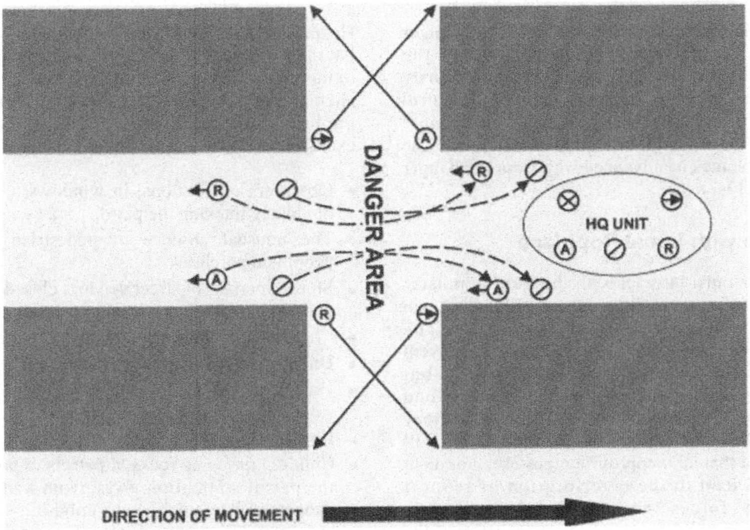

Figure 13-6. A&S Elements Move Through.

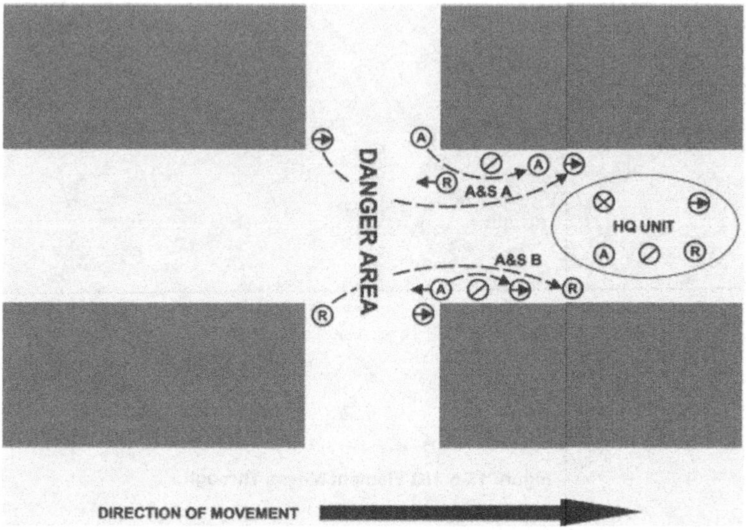

Figure 13-7. Patrol Resumes Movement Along Intended Route.

Once the trail A&S team members cross the danger area, they take up rear security and cover the movements of the near and far side flank security teams as they return to their positions in the patrol formation. The near side security team should collapse back first, followed by their far side counterparts. The patrol then resumes its advance away from the danger area (see fig. 13-7).

Interaction with Local Populace

Urban patrols must interact with the local populace. Patrols are at first a novelty to the civilians but can quickly become an unwelcome intrusion. The movement of the patrol must be fast enough to prevent the enemy from massing their fires upon it, but deliberate enough to ensure adequate security and mutual support. Patrol members must realize that they are usually the only Marines the local populace will encounter and that an inappropriate gesture, comment or act could lead to the deterioration of rapport between U.S. forces and the general population. Marines must remember that the vast majority of the individuals with whom they come in contact will be noncombatants attempting to survive in trying political, economic, and social situations.

Hostile incidents often seem to occur spontaneously, but there are usually indications that can alert Marines to imminent danger. The most obvious are the sudden alteration of normal routines, patterns, and attitudes of the local populace or other unusual activity. Some examples include—

- Observers on rooftops, in windows, etc., who are obviously tracking the patrol.
- The unusual absence of pedestrian traffic and people on porches.
- Stores, markets or street vendors closed suddenly or without explanation.
- Changes in civilian attitude toward patrol members.
- Unknown individuals or vehicles in the patrol area.
- Unfamiliar vehicles parked in the patrol area (possible car bomb).
- Roadblocks.
- Children throwing rocks at patrols to possibly draw the patrol's attention away from a more serious danger, such as a deliberate ambush.
- Vehicles riding unusually low due to overloading (possibly ferrying people, weapons, explosives).
- Agitators trying to provoke an incident with patrol members.

- Absence of the usual stray dogs (dogs are adept at sensing danger and avoiding it).
- Anti-American graffiti suddenly appearing in the patrol area.
- Pictures of enemy leaders and martyrs posted in the patrol area.
- Civilian workers failing to appear at U.S. or friendly bases.
- Normal deliveries and pick-ups conducted late or early without reason.
- Sudden change of civilian sentiment in newspaper articles, radio broadcasts or other media.
- Women and children leaving to live elsewhere.

Immediate Actions Upon Enemy Contact

Reaction to Sniping

Snipings are often executed from a single firing point, but coordinated snipings delievered from multiple points are not uncommon.

Patrol element or team leaders should constantly try to identify likely firing points and anticipate their own reactions to a possible shooting. Normally, snipers in an urban environment have a detailed withdrawal plan. Once a patrol comes into contact with a sniper, the patrol leader must immediately assess the situation and maneuver his patrol accordingly. The patrol's mission, location, size, ROE, and location of the threat often determine whether the patrol will attempt to neutralize the targeted sniper. If the patrol leader decides to kill or capture the sniper, he uses planned and rehearsed immediate actions to maneuver and counter the sniper's assault. The goal is to kill the sniper or cut off his escape and capture him. There are three immediate reactions to neutralizing a sniper: initial contact, immediate follow-up, and subsequent follow-up.

Initial Contact. The initial contact is made when the sniper fires the first shot. The patrol must react immediately and positively to get behind the firing position in order to kill or capture the gunman. The period of contact ends when the gunman is killed or captured, or the patrol element or team leader on the scene ends it. The following technique is the same for both squad- and platoon-sized patrols:

- The element or team in contact attempts to identify the firing position and maneuvers designated marksman into position to return well-aimed and controlled fire. Other members of the patrol take up positions to cover the marksman's engagement.
- The patrol element or team leader in contact sends initial contact report to the patrol leader, who notifies higher headquarters.
- The element or team leader in contact determines appropriate cut-off positions and relays them to flanking elements or teams.
- The patrol element or team leader in contact continues to observe the firing point, but does not enter it due to the possibility of booby traps. Flank elements or teams set up along likely escape routes.
- The incident ends when either the sniper ceases fire or is neutralized.

Immediate Follow-Up. Regardless of the fate of the gunman, isolation of the firing point is necessary to prevent reinforcement and preserve forensic evidence (scent, spent casings, etc.). If not under fire, members of the patrol element or team cordon off the area surrounding the firing point. Flanking elements or teams maintain their positions and prevent civilians from entering the area. The patrol leader moves to link-up with the element or team in contact (if not his own), and makes an estimate of the situation. The patrol leader sends a SPOTREP to higher headquarters.

Subsequent Follow-Up. The aim of the subsequent follow-up is to use follow-on forces to clear the building of remaining resistance or to obtain evidence that can be used to capture the gunman. The patrol leader establishes a position where he can brief arriving units (reaction force commander, S-2 representative, EOD personnel, etc.). Once the arriving units have been briefed, recommendations are made to higher headquarters via radio. No one is allowed into the cordon without the patrol leader's approval.

Reaction to Becoming Decisively Engaged

If a patrol becomes decisively engaged from numerous firing positions, the following immediate action should be taken:

- All patrol members move to available cover and return accurate fire on identified firing points.
- The patrol leader assesses the situation and makes a decision to either request the reaction force or break contact.
- If the reaction force is requested, the patrol will maintain its position until the reaction force arrives. The patrol should use fire and maneuver to gain better tactical positioning and support the arrival of the reaction force. When the reaction force arrives,

its commander may decide to either clear occupied buildings or cover the patrol during its extract.

Reaction to Bomb Threat or Discovery

The use of command-detonated explosive devices is a common ambush tactic employed by a terrorist or insurgent in an urban environment. The appropriate response to a reported threat or an actual discovery is generally involves four steps (known as the four Cs)—

1. The patrol leader **CONFIRMS** the presence of the suspicious item.

2. Without touching or moving anything suspicious, patrol elements or teams **CLEAR** the immediate danger area to a minimum of 100 meters. The area is cleared from the suspected device outward, inform civilians as to the reason for evacuation.

3. A&S teams establish a **CORDON** to secure the cleared area. Avenues of approach are cordoned off to keep people out and to protect EOD or engineer personnel clearing the device. The assistant patrol leader acts as the cordon commander and informs the patrol leader when the cordon is secure. An effective cordon technique is to tape off the area with engineer tape, creating both a physical and psychological boundary.

4. **CONTROL** of the area is maintained throughout the bomb clearing operation by the patrol leader. The patrol leader sends a report to a superior concerning details of the device (if known) and the area affected. The patrol leader coordinates with arriving personnel (EOD, engineers, etc.). The patrol leader maintains communication with the assistant patrol leader and keeps the Marines informed of the progress of the clearing operation.

Reaction to a Bomb Detonation

Bombs may be used by an insurgent as a means of initiating an ambush on mounted or dismounted patrols, in which case the actions for decisive engagement apply. Immediate action in response to an isolated explosion is similar to that used in reaction to a sniping and breaks down into the same three phases:

Initial Contact. The patrol leader attempts to identify the likely initiation point and sends an initial contact report to higher headquarters. If the bomb was command-detonated, the patrol leader sends his A&S teams deep to cut off the bombers' escape routes. Any casualties are moved a minimum of 100 meters from explosion and out of the line of sight to it.

Immediate Follow-Up. The A&S teams may need to maneuver to positions behind the likely initiation point to cut off escape. Once in position, personnel checks are conducted and any suspects are detained. The patrol leader coordinates requests for required support (MEDEVAC, reaction force, etc.)

Subsequent Follow-Up. Due to the possibility of secondary detonations, the four Cs (confirm, clear, cordon, and control) can be conducted as in reaction to a bomb discovery or bomb threat.

Civil Disturbances

Urban patrols must prepare to react to spontaneous aggression by the local populace. In many cases, civil disturbances are organized by the enemy to draw dismounted patrols into a targeted area, or to distract them from enemy activity occurring elsewhere. Civil disturbances are generally divided into two categories: minor aggressive actions, and full-scale rioting.

Minor aggressive actions are activities characterized by rock-throwing or use of devices such as Molotov cocktails and may either be directed at the patrol or take place between different ethnic factions of the population. Minor aggressive actions are normally spontaneous in nature and may have minimal or limited objectives for the insurgents.

Full-scale rioting events are usually in response to another major event or incident that may enflame the populace. Full-scale riots are well-planned and orchestrated, with clear objectives or targets in mind. At times, patrols will need to attempt to maintain control of a civil disturbance situation; however, dismounted and mobile small unit patrols should generally avoid potential flashpoints. Procedures to handle civil disturbances are as follows:

- The patrol leader reports the incident to headquarters and attempts to diffuse the crowd by talking to crowd leaders.
- If the patrol leader determines the size of the disturbance is too large for the force to handle, the patrol should move away from the disturbance to a safer, more remote covered area and occupy positions to observe and report the situation to higher headquarters. To prevent the patrol from being pursued by the crowd, the patrol should move quickly and change direction, often at road junctions, to gain distance from it.

Patrol members should maintain dispersion to create a more difficult target. They should face the crowd at all times to see and avoid any projectiles thrown. Individual self-discipline must be maintained throughout the disturbance. Marines charging into the crowd or throwing objects back at the crowd will only worsen the situation. If pursued or trapped, the patrol leader may consider using riot control agents (combat support, pepper gas, etc.) to disperse or slow a crowd's movement. The use of riot control agents must be authorized under the established ROE. If the patrol leader believes the situation is deteriorating beyond the patrol's ability to control it, the patrol leader should request the reaction force, which may be better equipped to handle a large riot or mob.

Break Contact

As with patrols in rural areas, the patrol leader may be forced to break contact as a result of decisive engagement with the enemy. On the basis of his estimate of the situation, the patrol leader will normally break contact in one of the following ways:

- As a patrol, with elements providing cover for movement as defined by clock direction and distance.
- As individual units/teams taking separate routes out of the area, then linking up at a designated rally point a safe distance away from the engagement.

As in any contact with enemy forces, smoke may be employed to screen movement. Fire support agencies can be utilized to suppress targets; riot control agents can be employed to disrupt enemy movement.

Reentry of Friendly Lines

The reentry of a dismounted patrol into an urban patrol base is no different from that of a patrol conducted in a rural area. The same planning considerations and control methods apply.

Missions Related to Urban Patrolling

House Calls

House calls missions are usually part of a coordinated effort to collect information within the area of operations. They involve obtaining up-to-date information on particular houses and occupants. When possible, local police should accompany patrols to do the actual talking to the occupants. If this is not possible, a technique that may be used by an urban patrol is—

- A&S elements move to provide cover around the target house.
- The headquarters element provides security just outside the house.
- The patrol leader and one Marine for security enter the house, if invited. If not invited, they talk to the occupants on the doorstep. Patience and tact are required in requesting information. An interpreter should be present when language differences exist.

Vehicle Checkpoints

Urban terrorists or insurgents commonly use vehicles to transport personnel, weapons, explosives, and equipment. Civilian vehicles are often used for these purposes, creating the requirement to check as many vehicles as possible. While permanent, fortified checkpoints may be conducted along approaches into an urbanized area, dismounted patrols can be employed to establish hasty vehicle checkpoints to stop vehicles and to keep the enemy off-guard. The two common types of vehicle checkpoints patrols establish are hasty and deliberate.

Hasty checkpoints are deployed anywhere based upon the decision of the patrol leader. Patrols must not set patterns through the frequent use of the same sites.

Deliberate checkpoints are tasked by higher headquarters to achieve a specific purpose. Time and locations are carefully considered to avoid setting patterns.

The general layout for a squad-sized, two-way dismounted checkpoint is depicted in figure 13-8 on page 13-20. The technique is as follows:

- The A&S teams are positioned stealthily in blocking positions on both sides of the road.
- Both the patrol leader and assistant patrol leader act as "talkers" for each direction of traffic (with local police or interrogator or translator Marines acting as interpreters) while a two-man team from the headquarters element physically searches the vehicles.
- Obstacles or parked vehicles may be employed to create a staggered roadblock in center of the checkpoint to slow approaching vehicles.
- The checkpoint location should be sited so that approaching vehicles cannot see it until they have passed a security team, and they have no escape route then available.
- Signs announcing the checkpoint should be displayed a safe distance from the search area for safety to both drivers and Marines.
- Normally higher headquarters will issue criteria that determines which vehicles are searched, but random checks of cars should normally be made as

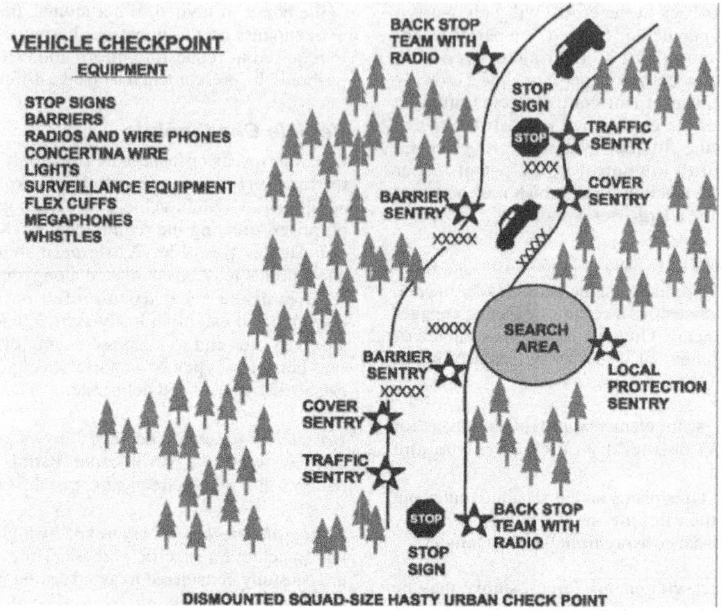

Figure 13-8. Dismounted Squad-sized Hasty Urban Vehicle Checkpoint.

well. While the vehicle is being searched the driver should accompany the searcher around the vehicle.

- Vehicle occupants should be made to exit the vehicle and then searched. Whenever possible, women should be searched by female police or female Marines.
- All patrol members should conduct themselves with courtesy and politeness. If nothing is found, an apology for the inconvenience is recommended.
- A hasty vehicle checkpoint should not be conducted any longer than 30 minutes for security reasons.
- The ROE should dictate whether or not action should be taken against vehicles that fail to stop at the checkpoint. Failure of a vehicle to stop does not automatically give authorization to fire.

Observation Posts

Urban observation posts are established to provide extended security, not only for patrol bases but also for patrols operating within the observation post's sector of observation. Observation posts can be established in conjunction with sniper operations and for providing overwatch for patrols operating within their sector of observation. Observation posts are nor-mally positioned on dominating terrain or in buildings outside the patrol base itself.

Insertion to the observation posts and conduct of observation may be either overt or clandestine in nature. Overt observation posts usually will be hardened positions to increase security. A patrol provides cover while the observation post is being inserted. A clandestine observation post relies on stealth of insertion and occupation for protection. It is normally positioned in abandoned buildings to cover sectors of observation that overt observation posts cannot. Because of their nature, clandestine observation posts are difficult to successfully establish and should not be manned for an extended period of time. Orders establishing observation posts (and patrols) must address the method of extraction as well as actions upon compromise/attack.

Cordon and Search

The cordon and search mission involves isolating a predesignated area by cordoning it off and systematically searching for enemy personnel, weapons, supplies, explosives or communications equipment. While large-scale cordon and search operations are planned and rehearsed in advance and normally entail extensive coordination with local law enforcement agencies, a squad-sized urban patrol may often conduct a cordon and search of a point target—searching one house or building identified by intelligence as a possible weapons cache.

The basic principle of a search of a populated area is to conduct it with limited inconvenience to the population. The populace may be inconvenienced to the point where they will discourage urban guerillas or insurgents from remaining in the area, but not to the point that they will assist the enemy as a result of the search.

Upon receiving intelligence that warrants the searching of a building or a specific tasking from higher headquarters, A&S elements of the patrol move to establish an inner cordon around the target building to seal it off, with the primary intent of preventing movement out of the targeted building.

On order, the designated reaction force deploys to establish an outer cordon, oriented outward some distance from the inner cordon and covering routes leading into the area in order to prevent outside interference/reinforcement. The reaction force maintains a reserve to reinforce either cordon or react to unfolding events (civil disturbance in response to the operation).

Once the cordons have been established, the patrol leader, with the assistance of local police or interpreters, informs the local populace that a building is about to be searched, that a house curfew is in effect (if permitted by higher headquarters), and that all occupants should remain indoors. Occupants of the target house are instructed to gather at a central location to stay out of the way of the search party.

The headquarters element, having linked up with any required assistance (explosive ordnance disposal (EOD), ITT, etc.) now acts as the search party and accompanies local police. A female searcher should be included in the party, if necessary.

Occupants are searched and screened first for possible enemy personnel. Apprehended persons are evacuated as soon as possible.

The head of the household should accompany the search party throughout the operation to be able to counter incriminating evidence and possible accusations of theft and looting against Marines. If possible, a prominent member of the local community should act as a witness.

Buildings are best searched from top to bottom. Ideally, the search is conducted with the assistance of combat engineers using mine detectors to locate hidden arms and ammunition.

If the targeted building is empty or the occupant refuses entry, it may be necessary to forcefully enter the premises to conduct a search of the dwelling. If an unoccupied house containing property is searched, arrangements should be made with the local community to secure it until its occupants return. Unnecessary force and damage to property should be avoided during the search.

Motorized Urban Patrols

The advantages of a motorized urban patrols is their ability to capitalize on the speed, mobility, and protection offered by various vehicles. They may be motorized, mechanized or armored vehicles or a combination. Generally, motorized urban patrols possess greater combat power than dismounted patrols and can cover larger areas faster than dismounted patrols.

The disadvantages of motorized urban patrols is that they are restricted to roads and are vulnerable to ambush by the enemy. They are also restricted in their ability to interact with the local populace.

Motorized patrols are generally organized in the same manner as dismounted patrols (see fig. 13-9). Unit integrity is maintained when assigning personnel to specific vehicles.

The urban patrolling principles apply to motorized patrols in much the same manner as dismounted patrols. Mutual support and depth are achieved by maintaining constant observation between vehicles and coordinating support with any dismounted patrols in the area. All-around security is achieved through the use of constant observation as well as the vehicle's mobility and firepower. Positive communications between units or teams are maintained through vehicle radios. Patrol routes and speeds are varied to promote deception or pattern avoidance.

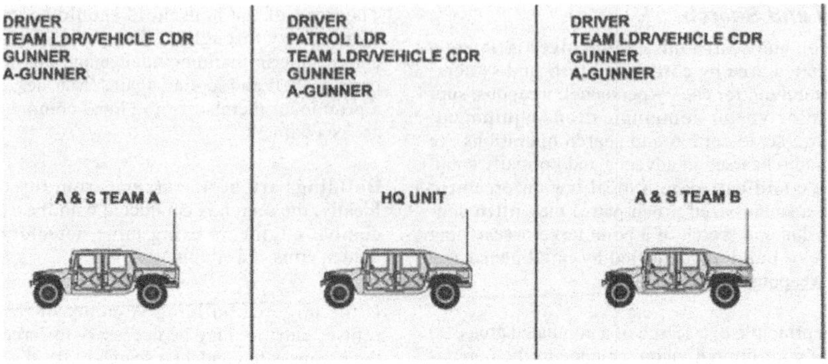

Figure 13-9. Organization of a Squad-sized Motorized Patrol.

The canalizing nature of streets and alleys force vehicular patrols to use a traveling overwatch movement technique to reduce vulnerability to ambushes. All vehicles travel at a moderate rate of speed with the lead vehicle stopping only to investigate potential danger areas. If vehicles must stop in danger areas, designated crew members' will dismount to provide security. The gunner will remain at the ready and in the turret while the driver remains in the driver's seat with the vehicle running.

Vehicles should move at a high rate of speed only when responding to an incident. At all other times, vehicle speed should be between 15 to 20 mph to allow for quick reaction and good observation. Distances between vehicles should be approximately 50 meters (one half to one city block) or such that visual contact and mutual support are ensured. Particular care is taken at major road junctions and other danger areas to ensure individual vehicles do not become isolated.

Vehicles with doors removed generally enhance observation and overall security, yet expose Marines to thrown objects, theft and concealment.

CHAPTER 14. INFORMATION AND REPORTS

It is necessary that patrol leaders and all patrol members be trained in observing and reporting their observations accurately. The leader of a patrol should have all members of the patrol immediately signal or report any information obtained. These reports should not be restricted to information about the enemy, but should also include information about the terrain, such as newly discovered roads, trails, swamps, and streams. The leader includes all information in the report to the officer dispatching the patrol. (Refer to MCRP 2-15.3, *Reconnaissance Reports Guide,* for detailed report formats.)

14001. REPORTING

The officer dispatching the patrol instructs the patrol leader on whether and when messages are to be sent back during the patrol and what communication means to use. Messages may be oral or written. They must be accurate, clear, and complete. Every message should answer the question what, where, and when. For a detailed discussion on reporting, see paragraph 6002.

Verbal Messages

A patrol leader sending a verbal message should make it simple, brief, and avoid using numbers and names. The messenger should accurately repeat the message back to the patrol leader before leaving.

Written Messages

In preparing written messages, the patrol leader must distinguish between fact and opinion. Information about the enemy should include: strength; armament and equipment; actions; location and direction of movement; unit destination, if known; time enemy was observed; and the patrol's location when the observation was made. Use of an overlay or sketch may often simplify the message.

Messengers

A messenger team is given exact instructions as to where to deliver the message and the route to take. Any information obtained along the route should be reported at the time the message is delivered. Messengers must be given all practical assistance. If in danger of capture, the messenger immediately destroys the message.

Radio and Signals

If the patrol is provided with a radio, a definite radio schedule for checking in must be arranged before departure of the patrol. The patrol leader takes every precaution to ensure that codes and copies of messages are not captured by the enemy. If a close reconnaissance of enemy lines is required, the radio should be left in a concealed location at a safe distance from the enemy. Once a report is sent by radio, the patrol should immediately leave the area to avoid the possibility of detection by enemy locating devices. Pyrotechnics (flares, colored smoke, grenades) and air-ground panels may also be used by patrols for reporting information by a prearranged signal.

SALUTE Report

Information must be reported as quickly, accurately, and as completely as possible. An established method to remember how and what to report about the enemy is to use the acronym SALUTE:

Size
Activity
Location
Unit
Time
Equipment

An example of such a report is: "Seven enemy soldiers, unit unknown, traveling SW, crossed road junction on BLACK RIDGE at 211300 August carrying one machine gun and one rocket launcher."

14002. CAPTURED ITEMS

Every patrol should make a practice of searching enemy casualties, prisoners, and installations first for booby traps, then for equipment, papers, maps, messages, orders, diaries, and codes. Search techniques should be practiced often in order to reduce the time exposed to potentially dangerous situations. Items found are collected by the patrol leader and turned in with the patrol report. The items found are marked as to time and place of capture. When possible, captured items should be linked to a specific prisoner who possessed the items or to the place where the items were found. When this is done, the enemy prisoner of war tag and item tag are marked accordingly. The patrol leader must impress upon the members of the patrol the importance of turning in all documents and equipment. Furthermore, the patrol leader must ensure all information gathered by the patrol is quickly disseminated.

14003. PRISONERS

A patrol normally does not capture prisoners unless required by the mission. If prisoners are taken, the "5S and T" rule applies. This memory aid stands for: search, segregate, silence, speed, safeguard, and tag.

Search. Prisoners are body-searched thoroughly for weapons and documents as soon as they have been captured. This search must include the helmet, body armor, and gas mask. These items are left with the prisoner for protection until the patrol is completed. Weapons, equipment, and documents are tagged and immediately sent to the patrol leader.

Segregate. Prisoners are segregated into isolated groups: officers, noncommissioned officers (NCOs), privates, deserters, and civilians. By segregating prisoners, it makes it more difficult for leaders to organize escapes and issue orders to subordinates.

Silence. Silence is essential. Do not allow prisoners to talk to each other.

Speed. Speed is required in getting prisoners to the commander who dispatched the patrol. Timely information secured from prisoners is essential.

Safeguard. Prisoners are safeguarded as they are moved. They are restrained, but not abused. If the patrol will soon reach friendly positions, prisoners are not given cigarettes, food or water until they have been questioned by interrogators. If the patrol will not return to a friendly position for a lengthy time, the prisoners are given food, water and medical aid.

Tag. Equipment and personal effects recovered from prisoners are tagged so that they may later be matched back to specific individuals.

14004. PATROL REPORT

Every patrol leader makes a report when the patrol returns. Unless otherwise directed, the report is made to the person ordering the patrol. If the situation permits, the report is written and supported by overlays and/or sketches. The patrol leader's report should be a complete account of everything of military importance observed or encountered by the patrol while on the assigned mission. It should include the following information:

- Size and composition of patrol.
- Tasks and purpose (mission).
- Time of departure.
- Time of return.
- Routes, out and back (show by sketch, azimuth, trace on map).
- Terrain (general description to include any man-made or natural obstacles and critical terrain features which, if occupied by either enemy or friendly forces, would allow them to control the surrounding area).
- Enemy (size, activity, location, unit, time, equipment).
- Any map corrections (show on map).
- Miscellaneous information not covered elsewhere in report.
- Results of enemy encounters.
- Condition of patrol, including disposition of any dead or wounded.
- Conclusion and recommendations.

14005. PATROL CRITIQUE

After the patrol has rested and eaten, the patrol leader should hold a critique. Constructive criticism is made. It is an excellent time to prepare for future patrols by going over lessons learned as a result of the patrol.

APPENDIX A. PATROL WARNING ORDER

The warning order is issued as soon as practical with all available information included to assist patrol members in preparation.

1. Situation. Friendly and enemy situation information necessary for initial preparation.

2. Mission. Statement of what the patrol is to accomplish, and the purpose for accomplishing it. When, how, and where will be discussed in the patrol order.

3. Execution

a. Task Organization. General patrol organization and assignment of responsibilities if known; otherwise, promulgate in the patrol order.

b. Tasks. Alert subordinate leaders to patrol tasks (e.g., stream crossing, helicopter rappelling, demolitions) requiring preparation prior to departure.

c. Coordinating Instructions

(1) Time schedule and location for individual preparation, rest, briefings, in-spection, rehearsal, and departure.

(2) Time, place, uniform, and equipment for receiving the patrol order.

(3) Tasks for subordinate leaders to direct and supervise the initial preparation which may include drawing ammunition, rations, and special equipment; conducting immediate action drills or other necessary individual or unit training; meeting and briefing attachment personnel; reconnoitering the area for passage of lines; and coordinating with the necessary unit leaders.

(4) Preliminary guidance to specialists and key individuals regarding their roles and organization within the patrol.

4. Administration and Logistics

a. Individual uniform, equipment, weapons, and prescribed load of rations, water, and ammunition.

b. Crew-served weapons (if required) and guidance regarding distribution of weapons and ammunition during movement.

c. Special equipment requirements (wire cutters, demolitions, radios, flashlights, infrared equipment, mines, binoculars) and their distribution during movement.

d. Restricted or prohibited items.

5. Command and Signal

a. Designation of assistant patrol leader and his role in preparation.

b. Designation of navigators and radio operators (if required).

c. Brief outline of patrol leader's schedule for preparation, and where he can be reached.

APPENDIX B. PATROL ORDER

A patrol order follows a warning order. Any of the following subjects that have been addressed in the warning order may be omitted with the exception of the mission statement. A patrol order is more detailed than a 5-paragraph squad order in that a great deal of attention is given to individual duties.

The patrol order also provides orientation information, both enemy and friendly, that impacts the patrol and enemy forces. This information can include, but is not limited to, weather, terrain, visibility, NBC considerations, local population situation, terrain model and/or map orientation, and behavior.

1. Situation

   a. Enemy Forces

      (1) Composition, disposition, and strength are based on size, activity, location, unit, time, and equipment (SALUTE).

      (2) Capabilities and limitations to defend, reinforce, attack, withdraw, and delay (DRAW-D).

      (3) Enemy's most probable course of action.

   b. Friendly Forces

      (1) Mission of next higher unit (task and commander's intent).

      (2) Adjacent unit missions (task and intent). Also identify left, front, right, and rear.

   c. Attachments and Detachments. (date and time effective).

2. Mission. The mission is the task to be accomplished, and its purpose (who, what, where, when, and why). For patrols, specify if the mission or time has priority.

3. Execution

   a. Commander's Intent and Concept of Operations

      (1) Commander's intent.

      (2) The concept of operations tells the where, how, and who and lays out the patrol leader's general scheme of maneuver and fire support plan. It outlines the following:

(a) Task organization of the patrol.

(b) Movement to the objective area, to include navigation method.

(c) Actions in the objective area.

(d) The return movement, to include navigation method.

(e) Use of supporting forces (including illumination, if required).

b. Subordinate Element Missions. Subordinate element missions (task and purpose) are assigned to elements, teams, and individuals, as required.

c. Coordinating Instructions. This paragraph contains instructions common to two or more elements, coordinating details, and control measures applicable to the patrol as a whole. At a minimum, it includes—

(1) Time of assembly in the assembly area.

(2) Time of inspections and rehearsals (if not already conducted).

(3) Time of departure and estimated time of return.

(4) Location of departure and reentry of friendly lines and the actions associated with departure and reentry.

(5) Details on the primary and alternate routes to and from the objective area.

(6) Details on formations and order of movement.

(7) Rally points and actions at rally points.

(8) Final preparation position and actions at this position.

(9) Objective rally point and actions at this point.

(10) Actions at danger areas.

(11) Actions in the event of enemy contact.

(12) Details on actions in the objective area not covered elsewhere.

(13) Estimated time of patrol debriefing upon return.

4. Administration and Logistics

a. Changes/additions to uniform, equipment, and prescribed loads from that given in the warning order.

b. Instructions for handling wounded and prisoners.

5. Command and Signal

a. Command Relationships. Identify key leaders and chain of command.

b. Signal. Challenge and password, arm and hand signals, special signals, and radio frequencies and call signs.

Appendix C. Patrol Evaluation Checklist

This appendix provides a comprehensive checklist of critical patrolling steps, techniques, and procedures to aid unit leaders to critique patrol performance during training. Unit leaders may use the list as they observe the performance of a patrol to provide the patrol leader with a detailed analysis of the performance.

Warning Order	Yes	No
Did the PL ensure all patrol members were present before issuing the warning order?		
Did the PL issue a brief statement on the enemy situation?		
Did the PL issue a brief statement on the friendly situation?		
Did the PL state the mission in a clear, positive manner?		
Did the PL list all members of the patrol including attachments?		
Was the chain of command for the patrol covered fully?		
Were all members of the patrol asigned positions and duties in particular squads and teams?		
Were all the necessary individual duties assigned?		
Did the PL follow established principles in organizing the patrol into squads and teams?		
Was each patrol member assigned a particular weapon to carry on patrol?		
Did the PL list all special equipment needed to accomplish the mission?		
Was required special equipment assigned to the proper element to carry?		
Did the PL select uniform and equiment common to all based on METT-T?		
Coordination with Adjacent Units		
Did the PL coordinate with other patrols operating to the right and left?		
Was the route out and back coordinated?		
Was the time of departure and return cocordinated?		
Were call signs and frequencies coordinated?		
Was a signal for the FPF coordinated so as not to approach friendly lines during this time?		
Coordination With Front Line Units		
Did the coordinator pass the size of the patrol?		
Was the time of departure and return coordinated?		
Did the coordinator give a general area of oeprations for the patrol?		
Did the coordinator ask for information on known or suspected enemy positions and/or obstacles?		
Did the coordinator ask about information on the latest enemy activity?		
Did the coordinator ask for detailed information on friendly fire support available and the unit's barrier plan?		
Was the location of the IRP established and coordinated?		
Did the coordinator ask the forward unit to monitor their patrol frequency?		
Was the current challenge and password confirmed?		
Did the coordinator request that all information coordinated be passed on to any relieving unti?		
Patrol Order		
General		
Did the PL check to ensure all patrol members were present before issuing the patrol order?		

	Yes	No
Did the PL issue the patrol order in a forceful, confident manner?		
Did the PL make maximum use of available visual aids in issuing the patrol order (i.e., terrain model, sand table, map board, chalkboard)?		
Did the PL issue the patrol order in correct sequence?		
Did the PL issue the entire patrol order without allowing interruptions by patrol members?		
Did the PL adequately answer all questions asked by patrol members		
Did paragraph 1a (Enemy Situation) include—		
A weather forecast for the period of operation?		
A description of the terrain over which the patrol was to operate?		
Identification or description of enemy units known to bein the area of operations?		
Known locations of enemy units?		
Recent activity of enemy units?		
Stength of enemy units in the operating area?		
Did paragraph 1b (Friendly Situation) include—		
Mission of the next higher unit?		
Locations, missions, and planned actions of units on right and left?		
Fire support available to support the patrol (as per the coordination)?		
Missions and routes of other patrols operating in the immediate area?		
Did paragraph 1c (Commander's Intent) include—		
All views and ides of what the commander wants to be accomplished?		
The final result desired by the commander?		
Did paragraph 1c (Attachments and Detachments) include—		
All attachments to patrol and effective time of attachment?		
All detachments from patrol and effective time of detachment?		
Did paragraph 2 (Mission) include, at a minimum—		
Who was to conduct the patrol?		
What the patrol was tasked to do (i.e., conduct a point recon patrol)?		
Where the action was to take place?		
Why the action was to take place (i.e., the purpose)?		
Did paragraph 3 a (Commander's Intent and Concept of Operations) include—		
A complete concept of operations?		
A detailed description of the mission of elements?		
A detailed description of the mission of teams?		
A detailed description of the duties of specific individuals (i.e., navigator, compass man, pace m an, corpsman, APL)?		
Time of departure and time of return?		

	Yes	No
Type of formations and order of movement to be used?		
Description of the route and alternate routes(s) to include azimuths and distances between checkpoints as a minimum?		
Techniques to be used in the departure from friendly areas?		
Techniques to be used for the reentry into friendly areas?		
Location of the IRP and a tentative ORP, to include grid coordinates and recognizable terrain features?		
Method for designating and passing rally points?		
Actions to be taken at rally points if their use became necessary?		
Actions to be taken in the event of enemy contact?		
Actions to be taken at danger areas?		
Complete, detailed description of actions to be taken at the objective?		
Times and locations for rehearsals and the order of priority for rehearsals?		
Times and lcoations for inspections and the methods of conduct (to include uniform and equipment to be worn and/or carried)?		
Where, when, and by whom the debriefing is to be conducted?		
Did paragraph 4 (Administration and Logistics) include—		
Rations to be carried or reference to the warning order if there is no change?		
Arms and ammunition to be carried or reference to the warning order if there is no change?		
Uniform and equipment to be worn and carried or reference to the warning order if there is no change?		
Method for handling wounded or dead?		
Method for handling prisoners?		
Did paragraph 5a (Signal) include—		
Complete description of all signals, code words, frequencies, and call signs to be used within the patrol?		
Were signals adequate?		
Radio call signs to be used?		
Primary and alternate frequencies to be used?		
Call signs and frequencies to be used with other units in the area of oeprations (if applicable)?		
Required reports to be transmitted to higher headquarters, along with occasions for reporting?		
Code words and brevity codes to be used between the patrol and higher headquarters?		
Current challenge and password to be used in friendly held areas?		
Current challenge and password to be used forward of FEBA?		
Did paragraph 5b (Command) include—		
Chain of command or reference to the warning order if there is not change?		
Location of the PL during all stages of operation?		

	Yes	No
Location of the APL during all stages of operation?		
Inspection Rehearsal		
General		
Did the P L adequately use the allotted inspection time?		
Was the inspection conducted in an orderly manner?		
During the inspection, did the PL check the following:		
Completeness and correctness of uniform?		
Completeness of all equipment necessary to accomplish the mission?		
Operational condition of equipment?		
Did the PL question members to ensure they knew—		
The mission of the patrol?		
The concept of operations?		
Their individual duties and responsibilities?		
Chain of command and succession?		
Duties and responsibilities of key personnel?		
Did the PL utilize the rehearsal area to conform as much as possible to area of operations?		
Were the following major actions rehearsed:		
Actions at the objective?		
Actions at danger areas?		
Actions on enemy contacts?		
Departure and reentry of friendly lines?		
Did the PL critique each action after it had been rehearsed?		
Were interpatrol communications and control measures checked at rehearsal?		
Did the PL maintain control of the patrol during the rehearsal?		
Movement		
Did the PL use proper formations for movement?		
Did the PL adequately control the point team?		
Did the PL check the compass men?		
Was the PL aware of his position at all times?		
Was light and noise discipline enforced?		
Was the rate of movement appropriate for the mission?		
Did the PL make full use of pace?		
Did the PL make full use of count?		
Was proper security maintained during movement?		
Did the PL select and pass proper rallying points?		
Did the PL ensure that all members knew location of rallying points?		

	Yes	No
Did the PL properly use arm and hand signals?		
Did the PL recognize and halt the patrol a safe distance from a danger area?		
If the PL conducted a map check, did he conduct it properly (light discipline, security, etc.)		
Did designated individuals reconnoiter the far side of a danger area properly?		
Were support teams emplaced properly?		
Did the PL use a proper formation in crossing the danger area?		
Were reports made on enemy contact and at checkpoints?		
Actions on Enemy Contact		
Did the PL halt patrol a safe distance from a tentative ORP?		
Did the PL issue an adequate frag order before departing to look for a tentative ORP?		
Did the PL take appropriate personnel to look for the ORP?		
Did the PL select a suitable ORP?		
Did the PL secure the site adequately?		
Did the patrol occupy the ORP as stated in the patrol order or subsequent frag order?		
Did the patrol occupy the ORP in an orderly manner?		
Did the PL maintain control of the patrol during occupation of the ORP?		
Did the PL issue a satisfactory contingency plan to the APL before departing on a leader's recon?		
Did the PL take appropriate personnel on a leader's recon?		
Did the PL maintain or ensure that the patrol leader had communications with the patrol?		
Did the patrol avoid being detected by the enemy during the leader's recon?		
If contact was made, did the PL take appropriate action?		
Was security maintained during the leader's recon?		
Did the PL ensure the objective was kept under surveillance?		
Were appropriate orders given to surveillance teams?		
If a leader recon proved a tentative ORP unsuitable, did the PL move the ORP?		
Was a satisfactory leader's recon conducted?		
Was the ORP move conducted in a satisfactory and orderly manner?		
On return to the ORP, did the PL issue appropriate frag order and allow enough time for dissemination?		
Actions at Objective		
Did the PL issue a frag order for action at the objective?		
Was security emplaced prior to the teams departing from the ORP?		
Did the PL employ the terrain at the objective to the best advantage?		
Was stealth maintained while moving into the objective?		

	Yes	No
Did the Pl exercise positive control of squads, teams, and individuals at the objective?		
Were actions at the objective in accordance with the details outlined in the patrol order?		
Did the PL make use of supporting arms at the objective?		
Was the action at the objective successful?		
Was withdrawal from the objective accomplished quickly and orderly?		
Did units withdraw according to the details reflected in the patrol order?		
Was reorganization at the ORP completed in an expeditious and orderly manner?		
Did the patrol withdraw from the ORP quickly and quietly?		

Appendix D. Ambush Formations

Formations discussed here are identified by names which correspond to the general pattern formed on the ground by deployment of the assault element.

Line Formation

The assault element is deployed generally parallel to the route of movement of the enemy. This positions the assault element parallel to the long axis of the killing zone and subjects the target to heavy flanking fire. The size of the force which can be trapped in the killing zone is limited by the area the assault element can effectively cover. The enemy is trapped in the killing zone by natural obstacles, mines, demolitions, and direct fires. (See fig. D-l.)

A disadvantage of the line formation is the chance that lateral dispersion of the target may be too great for effective coverage. The line formation is appropriate in close terrain that restricts enemy maneuver, and in open terrain where one flank is protected by natural obstacles or can be protected by mines and demolitions. Similar obstacles can be placed between the attack force and the killing zone to provide protection from possible enemy counter attack. When an ambush is deployed in this manner, access lanes are left so that the enemy can be assaulted. (See fig. D-2.)

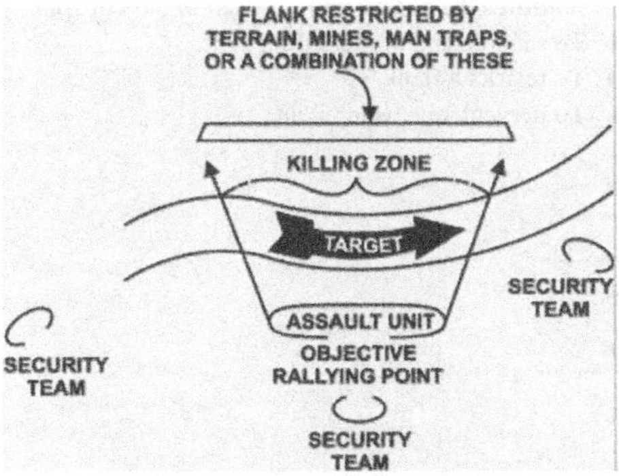

Figure D-1. Line Formation (Harassing or Destruction Ambush).

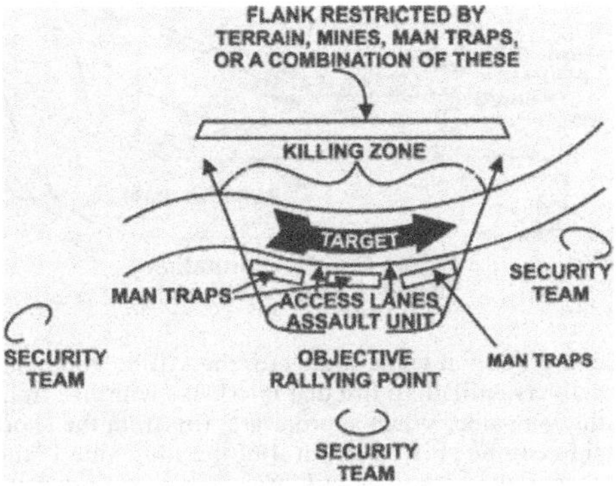

Figure D-2. Line Formation (Access Lanes for Assault of Target).

The main advantage of the line formation is its relative ease of control under all conditions of visibility.

L Formation

The "L" shaped formation is a variation of the line formation. This formation is very flexible because it can be established on a straight stretch of a trail or road (see fig. D-3) or at a sharp bend in a trail or a road (See fig. D-4 on page D-2).The long side of the assault element is parallel to the killing zone and delivers flanking fire. The short side of the attack force is at the

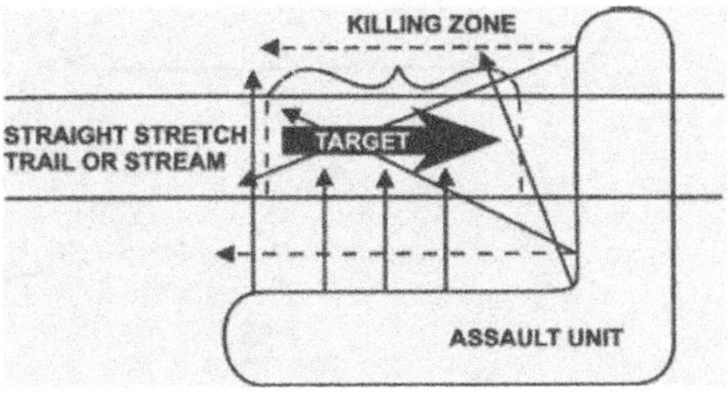

Figure D-3. "L" Formation (Destruction Ambush).

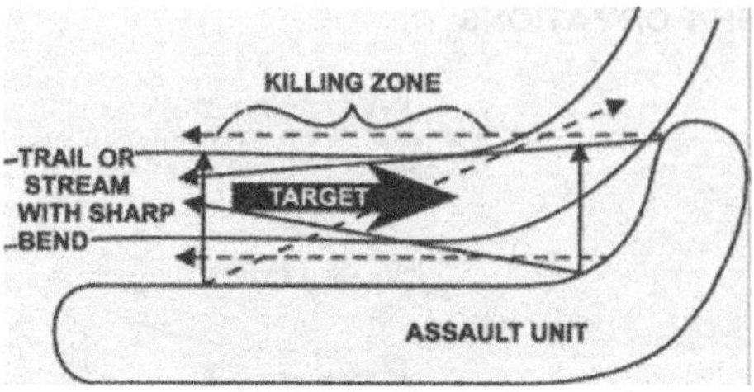

Figure D-4. "L" Formation (Bend of Trail or Stream).

end of, and at right angles to, the killing zone and delivers enfilading fire that interlocks with fire from the long side. When appropriate, fire from the short side can be shifted to parallel the long side if the enemy attempts to assault or escape in the opposite direction. In addition, the short side prevents escape and reinforcement. (See fig. D-5.)

"Z" Formation

The "Z" shaped formation is another variation of the line formation.

The assault element is deployed as in the "L" formation, but with an additional side so that the formation resembles a "Z". (See fig. D-6).

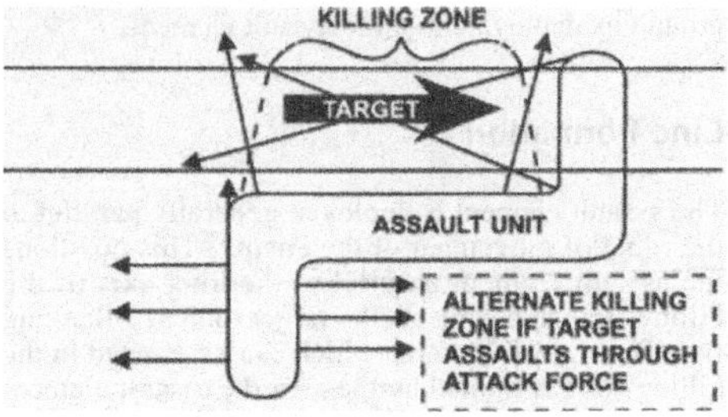

Figure D-6. "Z" Formation.

The additional wing may serve any of the following purposes:

- To engage an enemy force attempting to relieve or reinforce the enemy unit engaged in the kill zone.
- To seal the end of the killing zone.
- To restrict a flank.
- To prevent an envelopment.

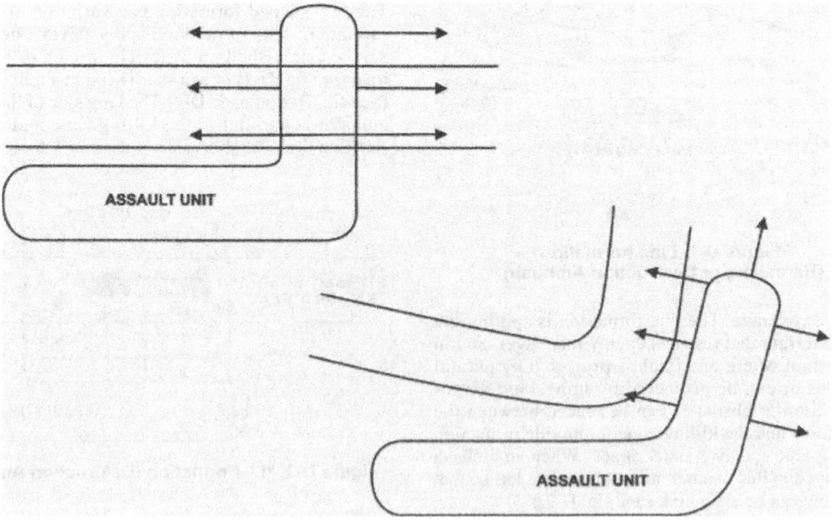

Figure D-5. "L" Formation.

"T" Formation

In the "T" shaped formation, the assault element is deployed across (perpendicular to) the enemy's route of movement so that its position forms the crossing of a "T" at the top. (See fig. D-7.)

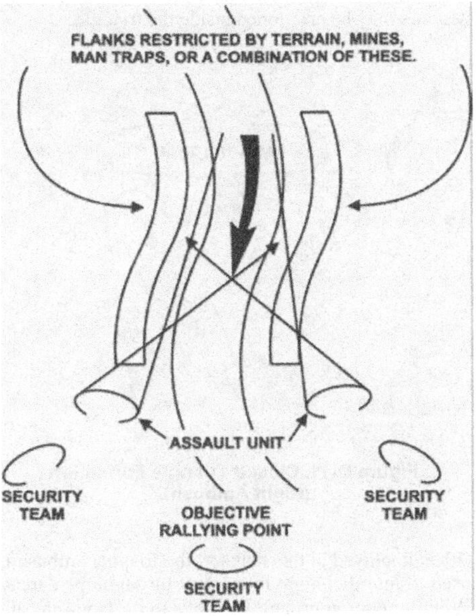

Figure D-7. "T" Formation.

This formation can be used day or night to establish an ambush to interdict movement through open areas that are hard to seal off.

A small force can use the "T" formation to harass, slow, and disorganize a larger force. When the lead elements of the enemy are engaged, they will normally attempt to maneuver right or left to close with the ambush. Mines and other obstacles placed to the flanks of the killing zones slow the enemy's movements and permit the ambush force to deliver heavy fire and withdraw without becoming decisively engaged.

The "T" formation can be used to interdict small groups attempting night movement across open areas. For example, the assault element is deployed along an avenue of approach with every second man facing the opposite direction. The attack of the enemy approaching from either direction requires only that every second man may shift to the opposite side of the formation. Each man fires only to his front and only when the enemy is at a very close range. Attack is by fire only and each man keeps the enemy under fire as long as it remains to his front.

If the enemy attempts to escape in either direction along the killing zone, each Marine takes the enemy under fire as the enemy comes into the Marine's sector of fire. The "T" formation is very effective at halting infiltration. But it has one chief disadvantage; there is a possibility that the ambush will engage a superior force at night while spread out. (See fig. D-8.)

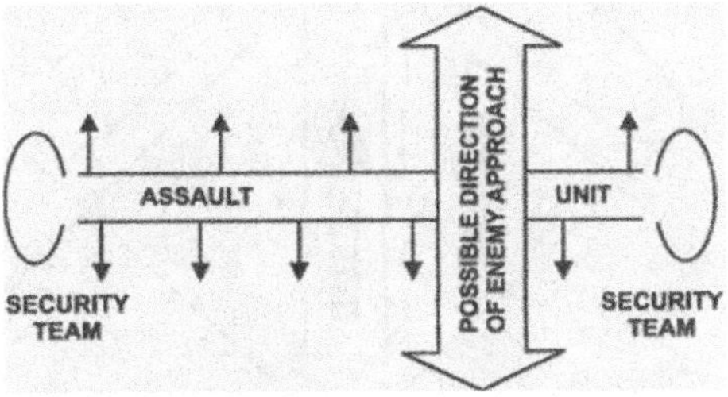

Figure D-8. "T" Formation (Target Approaching from Either Direction).

The "V" Formation

The "V" shaped formation is deployed along both sides of the enemy's route of movement so that it forms a "V"; care is taken to ensure that neither group (within the "V") fires into the other.

This formation subjects the enemy to both enfilading and interlocking fire. The "V" formation is best suited for fairly open terrain but can also be used in close terrain. When established in close terrain, the legs of the "V" close in as the lead element of the enemy force approaches the apex of the "V", and opens fire at a close range.

Here, even more than in open terrain, all movement and fire must be carefully coordinated and controlled to ensure that the fire of one wing does not endanger the other wing. The wider separation of forces makes this formation difficult to control, and there are few sites that favor its use. Its main advantage is that it is difficult for the enemy to detect the ambush until it is well into the killing zone. (See figs. D-9 and D-10.)

Triangle Formation

The triangle is a variation of the "V" formation and can be varied in three ways:

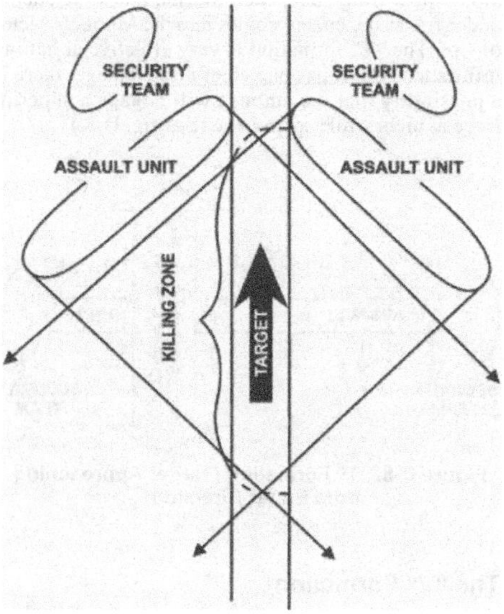

Figure D-9. "V" Formation (Open Terrain).

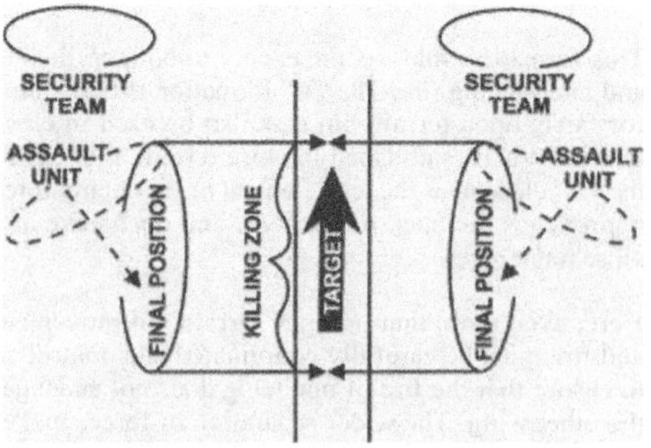

Figure D-10. "V" Formation (Close Terrain).

Closed Triangle Formation

(See fig. D-11.) The assault element is deployed in three teams, positioned so that they form a triangle (or closed "V"). An automatic weapon is placed at each point of the triangle and positioned so that it can be shifted quickly to interlock with either of the others. Men are positioned so that their sectors of fire overlap. Mortars may be positioned inside the triangle.

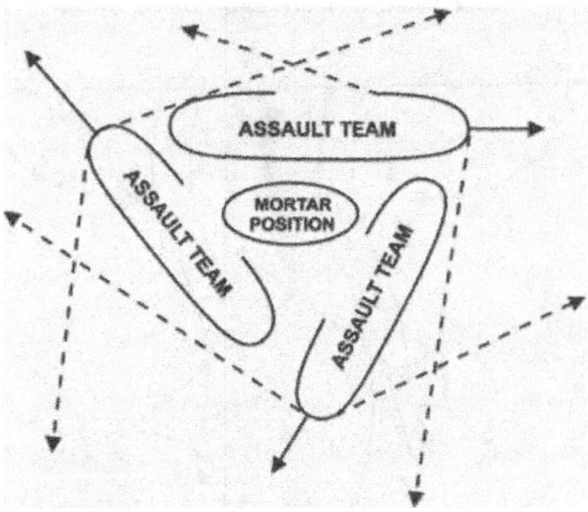

Figure D-11. Closed Triangle Formation (Night Ambush).

When deployed in this manner, the triangle ambush is used to interdict night movement through open areas. When enemy approach is likely to be from any direction, this formation provides all-around security, and security forces are deployed only when they can be positioned so that if detected by an approaching enemy, they will not compromise the ambush. Attack is by fire only, and the enemy is allowed to approach within close range before fire is initiated.

The advantages of the closed triangle formation are ease of control, all-around security, and the enemy can be brought under the fire of at least two automatic weapons, regardless of the direction they approach.

Disadvantages are that it requires a force of platoon size or larger to reduce the danger of being overrun by an unexpectedly large force and that one or more sides of the triangle may come under enfilade fire. The lack of dispersion, particularly at the points, increases the danger from enemy mortar fire.

Open Triangle Harassing Formation

This variation of the triangle ambush enables a small force to harass, slow, and inflict heavy casualties upon a large force without itself being decisively engaged. The assault element is deployed in three teams, positioned so that each team becomes a corner of a triangle containing the killing zone. (See fig. D-12.)

When the enemy enters the killing zone, the team to the enemy's front opens fire on the leading element. When the enemy counterattacks, the group withdraws and the team on the enemy's flank opens fire. When this team is attacked, the team to the opposite flank opens fire. This process is repeated until the enemy is pulled apart. Each team reoccupies its position, if possible, and continues to inflict the maximum damage possible without becoming decisively engaged.

Open Triangle Destruction Formation

The assault element is again deployed in three teams, positioned so that each team is a point of the triangle, 200 to 300 meters apart. The killing zone is the area within the triangle. The enemy is allowed to enter the killing zone; the nearest team attacks by fire. As the enemy attempts to maneuver or withdraw, the other teams open fire. One or more teams, as directed, assault or maneuver to envelop or destroy the enemy. (See fig. D-13 on page D-6.)

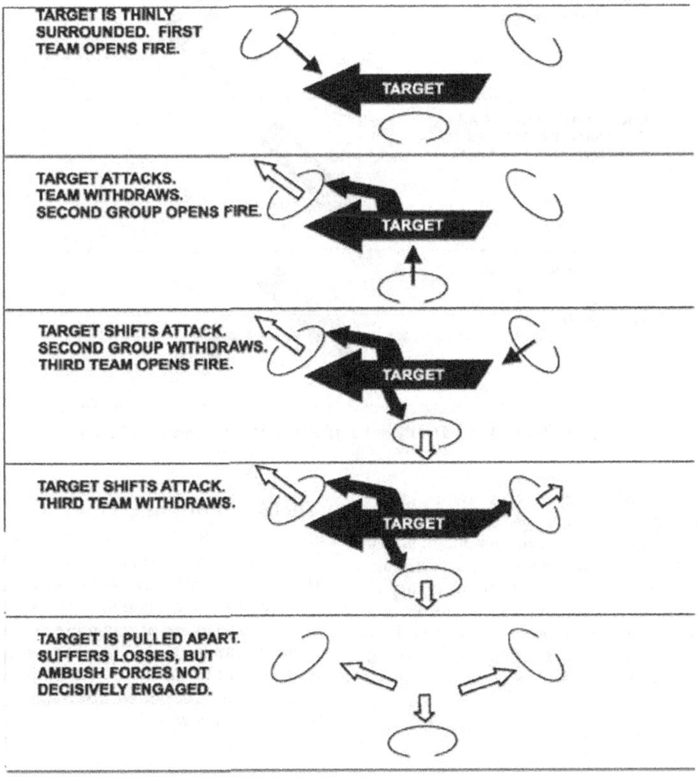

Figure D-12. Open Triangle Formation.

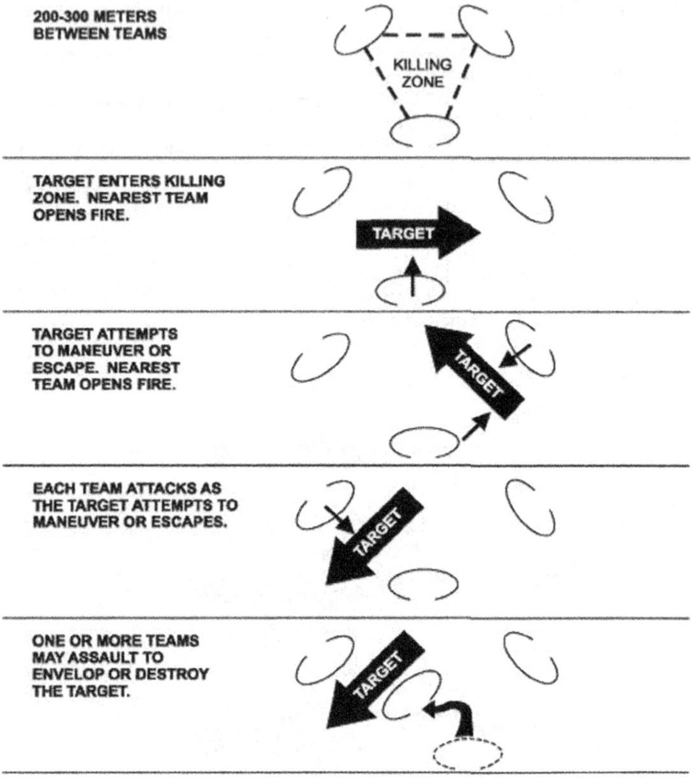

Figure D-13. Open Triangle Formation (Destruction Ambush).

This formation is suitable for platoon-size or larger ambush forces. A smaller force would be in too great a danger of being overrun. Another disadvantage is that control, in assaulting or maneuvering, is very difficult. Very close coordination and control are necessary to ensure that assaulting or maneuvering teams are not fired on by another team. The ambush site must be a fairly level open area that provides concealment around its border for the ambush force.

The Box Formation

The “box” formation is similar in purpose to the open triangle ambush. The assault element is deployed in four teams, positioned so that each team becomes a corner of a square or rectangle containing the killing zone. It can be used for a harassing or destruction ambush in the same manner as the variations of the open triangle formation. (See figs D-14 on page D-7 and D-15 on page D-8.)

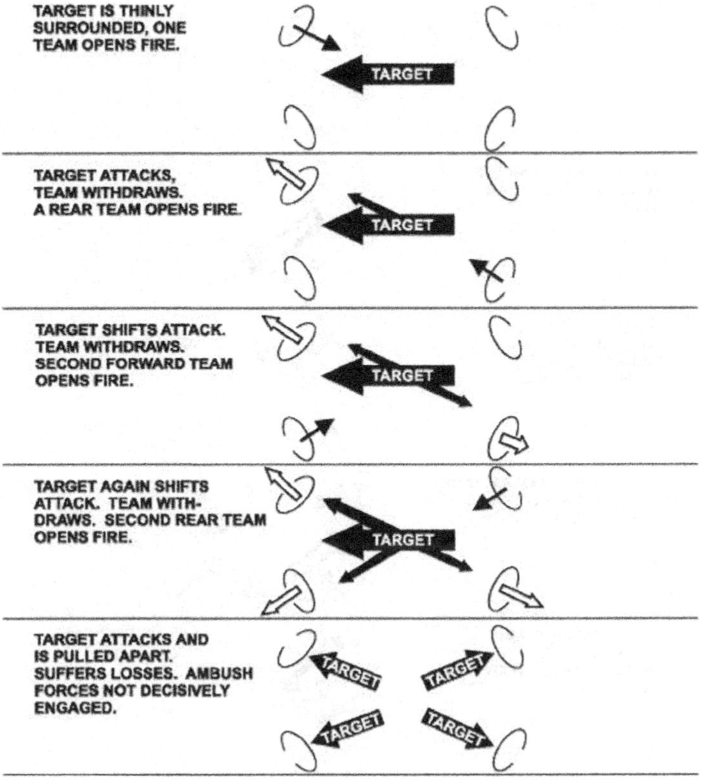

Figure D-14. Box Formation (Harrassing Ambush).

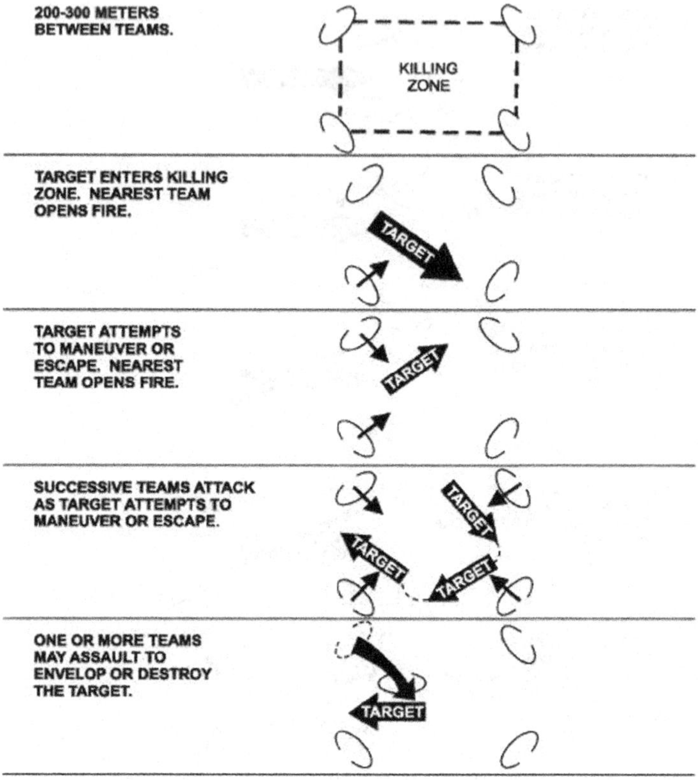

Figure D-15. Box Formation (Destruction Ambush).

Appendix E. Acronyms

A&S . . . . . assault and security

CIT . . . . . counterintelligence team

EOD . . . . . explosive ordnance disposal

ITT . . . . . interrogator-translator team

MAGTF . . . . . Marine air-ground task force
MEDEVAC . . . . . medical evacuation
METT-T . . . . . mission, enemy, terrain and weather, troops and support available

NBC . . . . . nuclear, biological, chemical
NCO . . . . . noncommissioned officer

OTL . . . . . overserver-target line

RCA . . . . . riot control agents
ROE . . . . . rules of engagement

SMAW . . . . . shoulder-launched multipurpose assault weapon
SMEAC . . . . . situation, mission, execution, administration and logistics, and command and signal
SOP . . . . . standing operating procedures
SPOTREP . . . . . spot report

TTP . . . . . tactics, techniques, and procedures

APPENDIX F. REFERENCES

Joint Publication (JP)

1-02	DOD Dictionary of Military and Associated Terms

Fleet Marine Force Manual (FMFM)

6-5	Marine Rifle Squad (proposed MCWP 3-11.2)

Marine Corps Warfighting Publications (MCWP)

3-15.3	Scout Sniping (under development)
3-35.3	Military Operations on Urbanized Terrain (MOUT)
3-41.2	Raids (under development)

Marine Corps Reference Publications (MCRP)

2-15.3A	Reconnaissance Patrol Leader's Planning Handbook (under development)
2-15.3B	Reconnaissance Reports Guide
3-02C	Water Survival Handbook (under development, currently exists as Fleet Marine Force Manual 0-13)
5-12A	Operational Terms and Graphics
5-12C	Marine Corps Supplement to the DOD Dictionary of Military and Associated Terms